Chemistry Workbook FOR DUMMIES®

by Peter J. Mikulecky, PhD, Katherine Brutlag, Michelle Rose Gilman, and Brian Peterson

WILEY

Wiley Publishing, Inc.

Chemistry Workbook For Dummies®

Published by
Wiley Publishing, Inc.
111 River St.
Hoboken, NJ 07030-5774
www.wiley.com

WILEY

About the Authors

Peter Mikulecky grew up in Milwaukee, an area of Wisconsin unique for its high human-to-cow ratio. After a breezy four-year tour in the Army, Peter earned a bachelor of science degree in biochemistry and molecular biology from the University of Wisconsin–Eau Claire and a PhD in biological chemistry from Indiana University. With science seething in his DNA, he sought to infect others with a sense of molecular wonderment. Having taught, tutored, and mentored in classroom and laboratory environments, Peter was happy to find a home at Fusion Learning Center and Fusion Academy. There, he enjoys convincing students that biology and chemistry are, in fact, fascinating journeys, not entirely designed to inflict pain on hapless teenagers. His military training occasionally aids him in this effort.

Katherine (Kate) Brutlag has been a full-fledged science dork since she picked up her first book on dinosaurs as a child. A native of Minnesota, Kate enjoys typical regional activities such as snow sports and cheese eating. Kate left Minnesota as a teen to study at Middlebury College in Vermont and graduated with a double major in physics and Japanese. Seeking to unite these two highly unrelated passions, she spent a year in Kyoto, Japan, on a Fulbright scholarship researching Japanese constellation lore. Kate was quickly drawn back to the pure sciences, however, and she discovered her love for education through her work at Fusion Academy, where she currently teaches upper-level sciences and Japanese.

Michelle Rose Gilman is most proud to be known as Noah's mom. A graduate of the University of South Florida, Michelle found her niche early, and at 19, she was working with emotionally disturbed and learning-disabled students in hospital settings. At 21, she made the trek to California, where she found her passion for helping teenagers become more successful in school and life. What started as a small tutoring business in the garage of her California home quickly expanded and grew to the point where traffic control was necessary on her residential street.

Today, Michelle is the founder and CEO of Fusion Learning Center and Fusion Academy, a private school and tutoring/test prep facility in Solana Beach, California, serving more than 2,000 students per year. She is the author of *ACT For Dummies, Pre-Calculus For Dummies, AP Biology For Dummies, AP Chemistry For Dummies, GRE For Dummies,* and other books on self-esteem, writing, and motivational topics. Michelle has overseen dozens of programs over the last 20 years, focusing on helping kids become healthy adults. She currently specializes in motivating the unmotivated adolescent, comforting their shell-shocked parents, and assisting her staff of 35 teachers.

Michelle lives by the following motto: There are people content with longing; I am not one of them.

Brian Peterson remembers a love for science going back to his own AP Biology class. At the University of San Diego, Brian majored in biology and minored in chemistry, with a pre-med emphasis. Before embarking to medical school, Brian took a young adult–professional detour and found himself at Fusion Learning Center and Fusion Academy, where he quickly discovered a love of teaching. Years later, he finds himself the science department head at Fusion and oversees a staff of 11 science teachers. Brian, also known as "Beeps" by his favorite students, encourages the love of science in his students by offering unique and innovative science curricula.

Dedication

We would like to dedicate this book to our families and friends who supported us during the writing process. Also, to all our students who motivate us to be better teachers by pushing us to find unique and fresh ways to reach them.

Authors' Acknowledgments

Thanks to Bill Gladstone from Waterside Productions for being an amazing agent and friend. Thanks to Georgette Beatty, our project editor, for her clear feedback and support. A special shout-out to our acquisitions editor, Lindsay Lefevere, who, for reasons unclear, seems to keep wanting to work with us. Acknowledgments also to our copy editor, Vicki Adang, and technical reviewer Michael Edwards.

Publisher's Acknowledgments

We're proud of this book; please send us your comments through our Dummies online registration form located at www.dummies.com/register/.

Some of the people who helped bring this book to market include the following:

Acquisitions, Editorial, and Media Development

Project Editor: Georgette Beatty

Acquisitions Editor: Lindsay Sandman Lefevere

Senior Copy Editor: Victoria M. Adang

Editorial Program Coordinator:
Erin Calligan Mooney

Technical Editor: Michael A. Edwards, PhD

Editorial Manager: Michelle Hacker

Editorial Assistants: Joe Niesen, Jennette ElNaggar

Cover Photo: Ty Milford

Cartoons: Rich Tennant (www.the5thwave.com)

Composition Services

Project Coordinator: Katherine Key

Layout and Graphics: Carl Byers, Melanee Habig, Joyce Haughey, Laura Pence

Proofreaders: Arielle Carole Mennelle

Indexer: Steve Rath

Publishing and Editorial for Consumer Dummies

Diane Graves Steele, Vice President and Publisher, Consumer Dummies

Joyce Pepple, Acquisitions Director, Consumer Dummies

Kristin A. Cocks, Product Development Director, Consumer Dummies

Michael Spring, Vice President and Publisher, Travel

Kelly Regan, Editorial Director, Travel

Publishing for Technology Dummies

Andy Cummings, Vice President and Publisher, Dummies Technology/General User

Composition Services

Gerry Fahey, Vice President of Production Services

Debbie Stailey, Director of Composition Services

Contents at a Glance

Table of Contents

Introduction

· ·

"The first essential in chemistry is that you should perform practical work and conduct experiments, for he who performs not practical work nor makes experiments will never attain the least degree of mastery."

—Jābir ibn Hayyān, 8th century

"One of the wonders of this world is that objects so small can have such consequences: Any visible lump of matter — even the merest speck — contains more atoms than there are stars in our galaxy."

—Peter W. Atkins, 20th century

Chemistry is at once practical and wondrous, humble and majestic. And, for someone studying it for the first time, chemistry can be tricky.

That's why we wrote this book. Chemistry is wondrous. Workbooks are practical. This is a chemistry workbook.

About This Book

When you're fixed in the thickets of stoichiometry or bogged down by buffered solutions, you've got little use for rapturous poetry about the atomic splendor of the universe. What you need is a little practical assistance. Subject by subject, problem by problem, this book extends a helping hand to pull you out of the thickets and bogs.

The topics covered in this book are those most often covered in a first course of chemistry. The focus is on problems — problems like those you may encounter in homework or on exams. We give you just enough theory to grasp the principles at work in the problems. Then we tackle example problems. Then *you* tackle practice problems.

This workbook is modular. You can pick and choose those chapters and types of problems that challenge you the most; you don't have to read this book cover to cover if you don't want to. If certain topics require you to know other topics in advance, we tell you so and point you in the right direction. Most importantly, worked-out solutions and explanations are provided for every problem.

Conventions Used in This Book

We provide the following conventions to guide you through this book:

- ✔ *Italics* highlight definitions, emphasize certain words, and point out variables in formulas.
- ✔ **Boldfaced** text points out key words in bulleted lists and actions to take in numbered lists.

Foolish Assumptions

We assume you have a basic facility with algebra and arithmetic. You should know how to solve simple equations for an unknown variable. You should know how to work with exponents and logarithms. That's about it for the math. At no point do we ask you to, say, consider the contradictions between the Schrödinger equation and stochastic wavefunction collapse.

We assume you're a high school or college student and have access to a secondary school-level (or higher) textbook in chemistry or some other basic primer, such as *Chemistry For Dummies* (written by John T. Moore, EdD, and published by Wiley). We present enough theory in this workbook for you to tackle the problems, but you'll benefit from a broader description of basic chemical concepts. That way, you'll more clearly understand how the various pieces of chemistry operate within a larger whole — you'll see the compound for the elements, so to speak.

We assume you don't like to waste time. Neither do we. Chemists in general aren't too fond of time-wasting, so if you're impatient for progress, you're already part-chemist at heart.

How This Book Is Organized

This workbook is divided into thematic parts. By no means is it absolutely necessary to consume all the chapters of a part in sequence, nor is it necessary to progress in a straight line from one part to the next. But it may be useful to do so, especially if you're starting from a place of Total Chemical Bewilderment (T.C.B.).

Part 1: Getting Cozy with Numbers, Atoms, and Elements

Chemists are part of a larger scientific enterprise, so they handle numbers with care and according to certain rules. The reasons for this meticulousness become clear as you consider the kinds of measurements chemists routinely make on very large numbers of particles. The most familiar of these kinds of chemical particles are atoms. This part covers some must-know material about numbers in chemistry, describes the basic structure of atoms, outlines how atoms belong to one or another variety of element, and explains how atoms interact within different states of matter.

Part II: Making and Remaking Compounds

Reactions are the dramatic deeds of chemistry. By reacting, atoms assemble into compounds, and compounds transform into other compounds. This part gives you the basic tools to describe the drama. We explain the basics of bonding and the system for naming compounds. We introduce you to the mole, to chemical equations, and to stoichiometry, simple concepts you'll use for the remainder of your chemical career — however long or brief.

Part III: Examining Changes in Terms of Energy

Chemistry is change. Change either happens or it doesn't. When it happens, change can occur rapidly or slowly. Busy and industrious as they are, chemists want to know whether their chemistry will happen and for how long. This part describes the kinds of changes that can occur in chemical systems and the kinds of systems — like, say, solutions — in which those changes occur. We cover the difference between equilibrium (will it happen?) and rate (how long will it take to happen?), and relate the two to differences in energy between states. Because chemistry transforms energy as well as matter, we explore some important ways chemists describe the changes in energy that drive chemical reactions.

Part IV: Swapping Charges

Charge is a big deal in chemistry. Charged particles are marquee players on the chemical playing field, and this part examines their playbook in greater detail. Acid-base reactions are vital chemical events that include the actions of charged particles such as hydrogen and hydroxide ions (H^+ and OH^- . . . you'll see). Oxidation-reduction (or "redox") reactions are another critical class of reactions that include transfers of electrons, the tiny, negatively charged particles that get most of the chemical action. Finally, we summarize nuclear chemistry, the special branch of chemistry that includes transformations of particles within the nucleus, the positively charged heart of an atom.

Part V: Going Organic

Because most practicing chemists are alive, it should come as no surprise that many of them are interested in the chemistry of life. The chemistry performed by living things is largely organic chemistry, or the chemistry of multicarbon compounds. Although organic chemistry is a central feature of living organisms, it's not limited to them. The energy and materials industries, for example, are chock-full of organic chemists. This part provides a concise overview of organic chemistry basics, highlighting simple structures and structural motifs, and surveying some important classes of organic molecules in biology.

Part VI: The Part of Tens

It's easy to get lost within a science that covers everything from subatomic particles to cellular phone batteries to atomic spectra from distant stars. When you grow dizzy with T.C.B., plant your feet on solid ground in the Part of Tens. This part is reassuringly succinct and practical, filled with the equations you need and helpful reminders about tricky details. Time spent in the Part of Tens is never wasted.

Icons Used in This Book

You'll find a selection of helpful icons nicely nestled along the margins of this workbook. Think of them as landmarks, familiar signposts to guide you as you cruise the highways of chemistry.

Within already pithy summaries of chemical concepts, passages marked by this icon represent the pithiest, must-know bits of information. You'll need to know this stuff to solve problems.

Sometimes there's an easy way and a hard way. This icon alerts you to passages intended to highlight an easier way. It's worth your while to linger for a moment. You may find yourself nodding quietly as you jot down a grateful note or two.

Chemistry may be a practical science, but it also has its pitfalls. This icon raises a red flag to direct your attention to easily made errors or other tricky items. Pay attention to this material to save yourself from needless frustration.

Within each section of a chapter, this icon announces "Here ends theory" and "Let the practice begin." Alongside the icon is an example problem that employs the very concept covered in that section. An answer and explanation accompany each practice problem.

Where to Go from Here

Where you go from here depends on your situation, your learning style, and your overall state of T.C.B.:

- If you're currently enrolled in a chemistry course, you may want to scan the Table of Contents to determine what material you've already covered in class. Do you recall any concepts that confused you? Try a few practice problems from these sections to assess your readiness for more advanced material.

- If you're brushing up on forgotten chemistry, it may be helpful to scan the chapters for example problems. As you read through them, you'll probably have one of two responses: 1) "Aaah, yes . . . I remember *that*." 2) "Oooh, no . . . I so do *not* remember that." Let your responses guide you.

- If you're just beginning a chemistry course, you can follow along in this workbook, using the practice problems to supplement your homework or as extra pre-exam practice. Alternately, you can use this workbook to preview material before you cover it in class, sort of like a spoonful of sugar to help the medicine go down.

Whatever your situation, be sure to make smart use of the practice problems, because they are the heart of the workbook. Work each practice problem completely — even if you suspect you're off-track — *before* you check your answer. If your answer was incorrect, be sure you reason through the provided answer and explanation so you understand *why* your answer was wrong. Then attempt the next problem.

Also, remember that the Cheat Sheet (the yellow tear-out card at the front of this book) and the Part of Tens are your friends. Most of the nitty-gritty stuff you need to work the more difficult problems is found in there.

Finally, rest assured that chemistry isn't alchemy. Mysterious as it may sometimes seem, chemistry is a practical, understandable pursuit. Chemistry is neither above you nor beyond you. It awaits your mastery.

"Science is, I believe, nothing but trained and organized common sense."

—Thomas H. Huxley, 19th century

Part I

Getting Cozy with Numbers, Atoms, and Elements

The 5th Wave   By Rich Tennant

"So what if you have a Ph.D. in chemistry? I used to have my own circus act."

In this part . . .

Chemistry explains things. The explanations include ideas like atoms and energy. A winding road of explanation stretches from the structure of a water molecule to the crash of a melting glacier in the Arctic. The road leads across the periodic table and is paved with numbers. In this part, we introduce you to the rules for handling numbers within chemistry and begin the exploration of the basic question of chemistry: How can a limited palette of elements paint the universe?

Chapter 1

Noting Numbers Scientifically

Chemistry is a science. This means that like any other kind of scientist, a chemist tests hypotheses by doing experiments. Better tests require more reliable measurements, and better measurements are those that have more accuracy and precision. This explains why chemists get so giggly and twitchy about high-tech instruments; those instruments make better measurements. How do chemists report their precious measurements? What's the difference between accuracy and precision in those measurements? How do chemists do math with measurements? These questions may not keep you awake at night, but knowing the answers to them will keep you from making embarrassing, rookie errors in chemistry. So we address them in this chapter.

Using Exponential and Scientific Notation to Report Measurements

Because chemistry concerns itself with ridiculously tiny things like atoms and molecules, chemists often find themselves dealing with extraordinarily small or extraordinarily large numbers. Numbers describing the distance between two atoms joined by a bond, for example, run in the ten-billionths of a meter. Numbers describing how many water molecules populate a drop of water run into the trillions of trillions.

To make working with such extreme numbers easier, chemists turn to *scientific notation*, which is a special kind of exponential notation. *Exponential notation* simply means writing a number in a way that includes exponents. Every number is written as the product of two numbers, a coefficient and a power of 10. In plain old exponential notation, a coefficient can be any value of a number multiplied by a power with a base of 10 (such as 10^4). But scientists have rules for coefficients in scientific notation. In scientific notation, a coefficient is always at least 1 and always less than 10 (such as 7, 3.48, or 6.0001).

To convert a very large or very small number to *scientific notation,* position a decimal point between the first and second digits. Count how many places you moved the decimal to the right or left, and that's the power of 10. If you moved the decimal to the left, the power is positive; to the right is negative. (You use the same process for exponential notation, but you can position the decimal anywhere.)

In scientific notation, the coefficients should be greater than 1 and less than 10, so look for the first digit other than 0.

To convert a number written in scientific notation back into decimal form, just multiply the coefficient by the accompanying power of 10.

Q. Convert 47,000 to scientific notation.

A. **47,000 = 4.7 × 10⁴.** First, imagine the number as a decimal:

47,000.00

Next, move the decimal between the first two digits:

4.7000

Then count how many positions to the left you moved the decimal (four, in this case), and write that as a power of 10: 4.7×10^4.

Q. Convert 0.007345 to scientific notation.

A. **0.007345 = 7.345 × 10⁻³.** First, move the decimal between the first two nonzero digits:

7.345

Then count how many positions to the right you moved the decimal (three, in this case), and write that as a power of 10: $0.007345 = 7.345 \times 10^{-3}$.

---

1. Convert 200,000 into scientific notation.

Solve It

2. Convert 80,736 into scientific notation.

Solve It

---

3. Convert 0.00002 into scientific notation.

Solve It

4. Convert 6.903×10^2 from scientific notation into decimal form.

Solve It

Multiplying and Dividing in Scientific Notation

A major benefit of presenting numbers in scientific notation is that it simplifies common arithmetic operations. (Another benefit is that, among the pocket-protector set, numbers with exponents just look way cooler.) The simplifying powers of scientific notation are most evident in multiplication and division. (As we describe in the next section, addition and subtraction benefit from exponential notation, but not necessarily from strict scientific notation.)

To multiply two numbers written in scientific notation, multiply the coefficients, and then add the exponents. To divide two numbers, simply divide the coefficients, and then subtract the exponent of the *denominator* (the bottom number) from the exponent of the *numerator* (the top number).

Q. Multiply, using the "shortcuts" of scientific notation: $(1.4 \times 10^2) \times (2.0 \times 10^{-5})$.

A. 2.8×10^{-3}. First, multiply the coefficients:

$1.4 \times 2.0 = 2.8$

Next, add the exponents of the powers of 10:

$10^2 \times 10^{-5} = 10^{2 + (-5)} = 10^{-3}$

Finally, join your new coefficient to your new power of 10:

2.8×10^{-3}

Q. Divide, using the "shortcuts" of scientific notation: $(3.6 \times 10^{-3}) / (1.8 \times 10^4)$.

A. 2.0×10^{-7}. First, divide the coefficients:

$3.6 / 1.8 = 2.0$

Next, subtract the exponent of the denominator from the exponent of the numerator:

$10^{-3} / 10^4 = 10^{-3-4} = 10^{-7}$

Then, join your new coefficient to your new power of 10:

2.0×10^{-7}

5. Multiply $(2.2 \times 10^9) \times (5.0 \times 10^{-4})$.

Solve It

6. Divide $(9.3 \times 10^{-5}) / (3.1 \times 10^2)$.

Solve It

7. Using scientific notation, multiply 52×0.035.

Solve It

8. Using scientific notation, divide $0.00809 / 20.3$.

Solve It

Using Exponential Notation to Add and Subtract

Addition or subtraction gets easier when your numbers are expressed as coefficients of identical powers of 10. To wrestle your numbers into this form, you might need to use coefficients less than 1 or greater than 10. So, scientific notation is a bit too strict for addition and subtraction, but exponential notation still serves us well.

To add two numbers easily by using exponential notation, first express each number as a coefficient and a power of 10, making sure that 10 is raised to the same exponent in each number. Then add the coefficients. To subtract numbers in exponential notation, follow the same steps, but subtract the coefficients.

Q. Use exponential notation to add these numbers: $3,710 + 2.4 \times 10^2$.

A. **39.5×10^2.** First, convert both numbers to the same power of 10:

37.1×10^2 and 2.4×10^2

Next, add the coefficients:

$37.1 + 2.4 = 39.5$

Finally, join your new coefficient to the shared power of 10:

39.5×10^2

Q. Use exponential notation to do this subtraction: 0.0743 − 0.0022.

A. 7.21×10^{-2}. First, convert both numbers to the same power of 10:

7.43×10^{-2} and 0.22×10^{-2}

Next, subtract the coefficients:

$7.43 - 0.22 = 7.21$

Then join your new coefficient to the shared power of 10:

7.21×10^{-2}

---

9. Add $398 \times 10^{-6} + 147 \times 10^{-6}$.

Solve It

10. Subtract $7.685 \times 10^5 - 1.283 \times 10^5$.

Solve It

---

11. Using exponential notation, add $0.00206 + 0.0381$.

Solve It

12. Using exponential notation, subtract $9,352 - 431$.

Solve It

Distinguishing between Accuracy and Precision

Accuracy and precision . . . precision and accuracy . . . same thing, right? Chemists everywhere gasp in horror, reflexively clutching their pocket protectors — accuracy and precision are different!

- ✔ *Accuracy* describes how closely a measurement approaches an actual, true value.

- ✔ *Precision*, which we discuss more in the next section, describes how close repeated measurements are to one another, regardless of how close those measurements are to the actual value. The bigger the difference between the largest and smallest values of a repeated measurement, the less precision you have.

The two most common measurements related to accuracy are *error* and *percent error*.

- ✔ **Error** measures accuracy, the difference between a measured value and the actual value:

  Actual value – Measured value = Error

- ✔ **Percent error** compares error to the size of the thing being measured:

  |Error| / Actual value = Fraction error

  Fraction error × 100 = Percent error

Being off by 1 meter isn't such a big deal when measuring the altitude of a mountain, but it's a shameful amount of error when measuring the height of an individual mountain climber.

Q. A police officer uses a radar gun to clock a passing Ferrari at 131 miles per hour (mph). The Ferrari was really speeding at 127 mph. Calculate the error in the officer's measurement.

A. **–4 mph.** First, determine which value is the actual value and which is the measured value:

Actual value = 127 mph; measured value = 131 mph

Then calculate the error by subtracting the measured value from the actual value:

Error = 127 mph – 131 mph = –4 mph

Q. Calculate the percent error in the officer's measurement of the Ferrari's speed.

A. **3.15%.** First, divide the absolute value (the size, as a positive number) of the error by the actual value:

|–4 mph| / 127 mph = 4 mph / 127 mph = 0.0315

Next, multiply the result by 100 to obtain the percent error:

Percent error = 0.0315 × 100 = 3.15%

13. Two people, Reginald and Dagmar, measure their weight in the morning by using typical bathroom scales, instruments that are famously unreliable. The scale reports that Reginald weighs 237 pounds, though he actually weighs 256 pounds. Dagmar's scale reports her weight as 117 pounds, though she really weighs 129 pounds. Whose measurement incurred the greater error? Whose incurred a greater percent error?

Solve It

14. Two jewelers were asked to measure the mass of a gold nugget. The true mass of the nugget was 0.856 grams (g). Each jeweler took three measurements. The average of the three measurements was reported as the "official" measurement with the following results:

Jeweler A: 0.863g, 0.869g, 0.859g

Jeweler B: 0.875g, 0.834g, 0.858g

Which jeweler's official measurement was more accurate? Which jeweler's measurements were more precise? In each case, what was the error and percent error in the official measurement?

Solve It

Expressing Precision with Significant Figures

After you know how to express your numbers in scientific notation and how to distinguish precision from accuracy (we cover both topics earlier in this chapter), you can bask in the glory of a new skill: using scientific notation to express precision. The beauty of this system is that simply by looking at a measurement, you know just how precise that measurement is.

When you report a measurement, you should only include digits if you're really confident about their values. Including added digits in a measurement means something — it means that you really know what you're talking about — so we call the included digits *significant figures*. The more significant figures in a measurement, the more precise that measurement must be. The last significant figure in a measurement is the only figure that includes any uncertainty. Here are the rules for deciding what is and what isn't a significant figure:

- **Any nonzero digit is significant.** So, 6.42 seconds (s) contains three significant figures.

- **Zeros sandwiched between nonzero digits are significant.** So, 3.07s contains three significant figures.

- **Zeros on the left side of the first nonzero digit are *not* significant.** So, 0.0642s and 0.00307s each contain three significant figures.

- **When a number is greater than 1, all digits to the right of the decimal point are understood to be significant.** So, 1.76s has three significant figures, while 1.760s has four significant figures. The "6" is uncertain in the first measurement, but is certain in the second measurement.

✔ **When a number has no decimal point, any zeros after the last nonzero digit *may or may not* be significant.** So, in a measurement reported as 1,370s, you can't be certain if the "0" is a certain value, or if it's merely a placeholder.

Be a good chemist. Report your measurements in scientific notation to avoid such annoying ambiguities (see the earlier section, "Using Exponential and Scientific Notation to Report Measurements").

✔ **Numbers from *counting* (for example, 1 kangaroo, 2 kangaroos, 3 kangaroos . . .) or from *defined quantities* (that is to say, 60 seconds per 1 minute) are understood to have an unlimited number of significant figures;** in other words, these values are completely certain.

✔ **If a number is already written in scientific notation, then all the digits in the coefficient are significant, and none others.**

So, the number of significant figures you use in a reported measurement should be consistent with your certainty about that measurement. If you know your speedometer is routinely off by 5 miles per hour, then you have no business protesting to a policeman that you were only going 63.2 miles per hour.

Q. How many significant figures are in the following three measurements?

20,175 yards, 1.75×10^5 yards, 1.750×10^5 yards

A. **Five, three, and four significant figures, respectively.** In the first measurement,

all digits are nonzero, except for a 0 that is sandwiched between nonzero digits, which counts as significant. The second measurement contains only nonzero digits. The third measurement contains a 0, but that 0 is the final digit and to the right of the decimal point, and is therefore significant.

---

15. Modify the following three measurements so that each possesses the indicated number of significant figures (SF) and is expressed properly in scientific notation.

76.93×10^{-2} meters (1 SF), 0.0007693 meters (2 SF), 769.3 meters (3 SF)

Solve It

16. In chemistry, the potential error associated with a measurement is often reported alongside the measurement, as in: 793.4 ±0.2 grams. This report indicates that all digits are certain except the last, which may be off by as much as 0.2 grams in either direction. What, then, is wrong with the following reported measurements?

893.7 ±1 gram, 342 ±0.01 gram

Solve It

Doing Arithmetic with Significant Figures

Doing chemistry means making a lot of measurements. The point of spending a pile of money on cutting-edge instruments is to make really good, really precise measurements. After you've got yourself some measurements, you roll up your sleeves, hike up your pants, and do math with the measurements.

When doing that math, you need to follow some rules to make sure that your sums, differences, products, and quotients honestly reflect the amount of precision present in the original measurements. You can be honest (and avoid the skeptical jeers of surly chemists) by taking things one calculation at a time, following a few simple rules. One rule applies to addition and subtraction, and another rule applies to multiplication and division.

> ✔ **When adding or subtracting, round the sum or difference to the same number of decimal places as the measurement with the fewest decimal places.** Rounding like this is honest, because you acknowledge that your answer can't be any more precise than the least precise measurement that went into it.

> ✔ **When multiplying or dividing, round the product or quotient so that it has the same number of significant figures as the least precise measurement — the measurement with the fewest significant figures.**

Notice the difference between the two rules. When you add or subtract, you assign significant figures in the answer based on the number of decimal places in each original measurement. When you multiply or divide, you assign significant figures in the answer based on the total number of significant figures in each original measurement.

Caught up in the breathless drama of arithmetic, you may sometimes perform multi-step calculations that include addition, subtraction, multiplication, and division, all at once. No problem. Follow the normal order of operations, doing multiplication and division first, followed by addition and subtraction. At each step, follow the simple rules previously described, and then move on to the next step.

Q. Express the following sum with the proper number of significant figures:

35.7 miles + 634.38 miles + 0.97 miles = ?

A. **671.1 miles.** Adding the three values yields a raw sum of 671.05 miles. However, the 35.7 miles measurement extends only to the tenths place; the answer must therefore be rounded to the tenths place, from 671.05 to 671.1 miles.

Q. Express the following product with the proper number of significant figures:

27 feet × 13.45 feet = ?

A. 3.6×10^2 **feet^2.** Of the two measurements, one has two significant figures (27 feet) and the other has four significant figures (13.45 feet). The answer is therefore limited to two significant figures. The raw product, 363.15 feet^2, must be rounded. You could write 360 feet^2, but doing so implies that the final 0 is significant and not just a placeholder. For clarity, express the product in scientific notation, as 3.6×10^2 feet^2.

17. Express this difference using the appropriate number of significant figures:

127.379 seconds − 13.14 seconds + 1.2×10^{-1} seconds = ?

18. Express the answer to this calculation using the appropriate number of significant figures:

345.6 feet × (12 inches / 1 foot) = ?

19. Report the difference using the appropriate number of significant figures:

3.7×10^{-4} minutes − 0.009 minutes = ?

20. Express the answer to this multistep calculation using the appropriate number of significant figures:

87.95 feet × 0.277 feet + 5.02 feet − 1.348 feet / 10.0 feet = ?

Answers to Questions on Noting Numbers Scientifically

The following are the answers to the practice problems presented in this chapter.

1 2×10^5. Move the decimal point immediately after the 2 to create a coefficient between 1 and 10. Because this means moving the decimal point five places to the left, multiply the coefficient of 2 with the power 10^5.

2 8.0736×10^4. Move the decimal point immediately after the 8 to create a coefficient between 1 and 10. This involves moving the decimal point four places to the left, so multiply the coefficient of 8.0736 with the power 10^4.

3 2×10^{-5}. Move the decimal point immediately after the 2 to create a coefficient between 1 and 10. This means moving the decimal point five spaces to the right, so multiply the coefficient of 2 with the power 10^{-5}.

4   **690.3.** This question requires you to understand the meaning of scientific notation in order to reverse the number back into "regular" decimal form. Because 10^2 equals 100, multiply the coefficient 6.903 with 100. This moves the decimal point two spaces to the right.

5 1.1×10^6. The raw calculation yields 11×10^5, which converts to the given answer when expressed in scientific notation.

6 3.0×10^{-7}. The ease of math with scientific notation shines through in this problem. Dividing the coefficients yields a coefficient quotient of 3.0, while dividing the powers yields a quotient of 10^{-7}. Marrying the two quotients produces the given answer, already in scientific notation.

7   **1.82.** First, convert each number to scientific notation: 5.2×10^1 and 3.5×10^{-2}. Next, multiply the coefficients: $5.2 \times 3.5 = 18.2$. Then add the exponents on the powers of 10: $10^{1 + (-2)} = 10^{-1}$. Finally, join the new coefficient with the new power: 18.2×10^{-1}. Expressed in scientific notation, this answer is $1.82 \times 10^0 = 1.82$.

8 3.99×10^{-4}. First, convert each number to scientific notation: 8.09×10^{-3} and 2.03×10^1. Then divide the coefficients: $8.09 / 2.03 = 3.99$. Next, subtract the exponent on the denominator from the exponent of the numerator to get the new power of 10: $10^{-3-1} = 10^{-4}$. Join the new coefficient with the new power: 3.99×10^{-4}. Finally, express gratitude that the answer is already conveniently expressed in scientific notation.

9 545×10^{-6}. Because the numbers are each already expressed with identical powers of 10, you can simply add the coefficients: $398 + 147 = 545$. Then join the new coefficient with the original power of 10.

10 6.402×10^5. Because the numbers are each expressed with the same power of 10, you can simply subtract the coefficients: $7.685 - 1.283 = 6.402$. Then join the new coefficient with the original power of 10.

11 40.16×10^{-3} **(or an equivalent expression).** First, convert the numbers so they each use the same power of 10: 2.06×10^{-3} and 38.1×10^{-3}. Here, we used 10^{-3}, but you can use a different power, so long as the same power is used for each number. Next, add the coefficients: $2.06 + 38.1 = 40.16$. Finally, join the new coefficient with the shared power of 10.

12 89.21×10^2 **(or an equivalent expression).** First, convert the numbers so each uses the same power of 10: 93.52×10^2 and 4.31×10^2. Here, we picked 10^2, but any power is fine so long as the two numbers have the same power. Then subtract the coefficients: $93.52 - 4.31 = 89.21$. Finally, join the new coefficient with the shared power of 10.

13   **Reginald's measurement incurred the greater magnitude of error, while Dagmar's measurement incurred the greater percent error.**

Reginald's scale reported with an error of 256 pounds – 237 pounds = 19 pounds. Dagmar's scale reported with an error of 129 pounds – 117 pounds = 12 pounds. Comparing the *magnitudes* of error, we see that 19 pounds > 12 pounds. However, Reginald's measurement had a percent error of 19 pounds / 256 pounds × 100 = 7.4%, while Dagmar's measurement had a percent error of 12 pounds / 129 pounds × 100 = 9.3%.

14 Jeweler A's "official" average measurement was 0.864g, while Jeweler B's official measurement was 0.856g; thus, **Jeweler B's official measurement is more *accurate* because it's closer to the actual value of 0.856g.**

However, **Jeweler A's measurements were more *precise*** because the differences between A's measurements were much smaller than the differences between B's measurements. Despite the fact that Jeweler B's average measurement was closer to the actual value, the *range* of his measurements (that is, the difference between the largest and the smallest measurements) was 0.041g. The range of Jeweler A's measurements was 0.010g.

This example shows how low precision measurements can yield highly accurate results through averaging of repeated measurements. In the case of Jeweler A, the error in the official measurement was 0.864g – 0.856g = 0.008g. The corresponding percent error was 0.008g / 0.856g × 100 = 0.9%. In the case of Jeweler B, the error in the official measurement was 0.856g – 0.856g = 0.000g. Accordingly, the percent error was 0%.

15 With the correct number of significant figures and expressed in scientific notation, the measurements should read as follows: 8×10^{-1} **meters, 7.7×10^{-4} meters, 7.69×10^{2} meters.**

16 **"893.7 ± 1 gram" is an improperly reported measurement because the reported value, 893.7, suggests that the measurement is certain to within a few tenths of a gram.** The reported error is known to be greater, at ±1 gram. The measurement should be reported as "894 ±1 gram."

"342 ±0.01 gram" is improperly reported because the reported value, 342, gives the impression that the measurement becomes uncertain at the level of grams. The reported error makes clear that uncertainty creeps into the measurement only at the level of hundredths of a gram. The measurement should be reported as "342.00 ±0.01 gram."

17 1.1436×10^{2} **seconds.** The trick here is remembering to convert all measurements to the same power of 10 before comparing decimal places for significant figures. Doing so reveals that 1.2×10^{-1} seconds goes to the hundredths of a second, despite the fact that the measurement contains only two significant figures. The raw calculation yields 114.359 seconds, which rounds properly to the hundredths place (taking significant figures into account) as 114.36 seconds, or 1.1436×10^{2} seconds in scientific notation.

18 4.147×10^{3} **inches.** Here, you must recall that defined quantities (1 foot is defined as 12 inches) have unlimited significant figures. So, our calculation is limited only by the number of significant figures in the 345.6 feet measurement. When you multiply 345.6 feet by 12 inches per foot, the feet cancel, leaving units of inches. The raw calculation yields 4,147.2 inches, which rounds properly to four significant figures as 4,147 inches, or 4.147×10^{3} inches in scientific notation.

19 -9×10^{-3} **minutes.** Here, it helps here to convert all measurements to the same power of 10 so you can more easily compare decimal places in order to assign the proper number of significant figures. Doing so reveals that 3.7×10^{-4} minutes goes to the hundred-thousandths of a minute, while 0.009 minutes goes to the thousandths of a minute. The raw calculation yields –0.00863 minutes, which rounds properly to the thousandths place (taking significant figures into account) as –0.009 minutes, or -9×10^{-3} minutes in scientific notation.

20 2.93×10^{1} **feet.** Following standard order of operations, this problem can be executed in two main steps, first performing multiplication and division, and then performing addition and subtraction.

Following the rules of significant figure math, the first step yields: 24.4 feet + 5.02 feet – 0.135 feet. Each product or quotient contains the same number of significant figures as the number in the calculation with the fewest number of significant figures.

The second step yields 29.3 feet, or 2.93×10^{1} feet in scientific notation. The final sum goes only to the tenths place, because the number in the calculation with the fewest decimal places went only to the tenths place.

Chapter 2

Using and Converting Units

Have you ever been asked for your height in centimeters, your weight in kilograms, or the speed limit in kilometers per hour? These measurements may seem a bit odd to those folks who are used to feet, pounds, and miles per hour, but the truth is that scientists sneer at feet, pounds, and miles. Because scientists around the globe constantly communicate numbers to each other, they prefer a highly systematic, standardized system. The *International System* of units, abbreviated *SI* from the French term *Système International,* is the unit system of choice in the scientific community.

You find in this chapter that the SI system is a very logical and well organized set of units. Despite what many of their hairstyles may imply, scientists love logic and order, so that's why SI is their system of choice.

As you work with SI units, try to develop a good sense for how big or small the various units are. Why? That way, as you're doing problems, you have a sense for whether your answer is reasonable.

Familiarizing Yourself with Base Units and Metric System Prefixes

The first step in mastering the SI system is to figure out the base units. Much like the atom, the SI base units are building blocks for more complicated units. In later sections of this chapter, you find out how more complicated units are built from the SI base units. The five SI base units that you need to do chemistry problems (as well as their familiar, non-SI counterparts) are given in Table 2-1.

Table 2-1	SI Base Units		
Measurement	**SI Unit**	**Symbol**	**Non-SI Unit**
Amount of a substance	Mole	mol	No non-SI unit
Length	Meter	m	Feet, inch, yard, mile
Mass	Kilogram	Kg	Pound
Temperature	Kelvin	K	Degree Celsius or Fahrenheit
Time	Second	s	Minute, hour

Chemists routinely measure quantities that run the gamut from very small (the size of an atom, for example) to extremely large (such as the number of particles in one mole). Nobody (not even chemists) likes dealing with scientific notation (which we cover in Chapter 1) if they don't have to. For these reasons, chemists often use metric system prefixes in lieu of scientific notation. For example, the size of the nucleus of an atom is roughly 1 *nano*meter across, which is a nicer way of saying 1×10^{-9} meters across. The most useful of these prefixes are given in Table 2-2.

Table 2-2	The Metric System Prefixes		
Prefix	**Symbol**	**Meaning**	**Example**
Kilo	k	10^3	$1 \text{ km} = 10^3\text{m}$
Deco	D	10^1	$1 \text{ Dm} = 10^1\text{m}$
Main Unit	varies	1	1m
Deci	d	10^{-1}	$1 \text{ dm} = 10^{-1}\text{m}$
Centi	c	10^{-2}	$1 \text{ cm} = 10^{-2}\text{m}$
Milli	m	10^{-3}	$1 \text{ mm} = 10^{-3}\text{m}$
Micro	μ	10^{-6}	$1 \text{ μm} = 10^{-6}\text{m}$
Nano	n	10^{-9}	$1 \text{ nm} = 10^{-9}\text{m}$

TIP

Feel free to refer to Table 2-2 as you do your problems. You may want to earmark this page because, after this chapter, we simply assume that you know how many meters are in one kilometer, how many grams are in one microgram, and so on.

EXAMPLE

Q. You measure a length to be 0.005m. How might this be better expressed using a metric system prefix?

A. **5 mm.** 0.005 is 5×10^{-3}m, or 5 mm.

1. How many nanometers are in 1 centimeter?

Solve It

2. If your lab partner has measured the mass of your sample to be 2,500g, how might you record this more nicely (without scientific notation) in your lab notebook using a metric system prefix?

Solve It

Building Derived Units from Base Units

Chemists aren't satisfied with measuring length, mass, temperature, and time alone. On the contrary, chemistry often deals in quantities. These kinds of quantities are expressed with *derived units,* which are built from combinations of base units.

- ✔ **Area (for example, catalytic surface):** Area = Length × Width and has units of length squared ($meter^2$, for example).

- ✔ **Volume (of a reaction vessel, for example):** You calculate volume by using the familiar formula: Volume = Length × Width × Height. Because length, width, and height are all length units, you end up with length × length × length, or a length cubed (for example, $meter^3$).

- ✔ **Density (of an unidentified substance):** Density, arguably the most important derived unit to a chemist, is built by using the basic formula, Density = Mass / Volume.

In the SI system, mass is measured in kilograms. The standard SI units for mass and length were chosen by the Scientific Powers That Be because many objects that you encounter in everyday life weigh between 1 and 100 kg and have dimensions on the order of 1 meter. Chemists, however, are most often concerned with very small masses and dimensions; in such cases, grams and centimeters are much more convenient. Therefore, the standard unit of density in chemistry is grams per cubic centimeter (g/cm^3), rather than kilograms per cubic meter.

The cubic centimeter is exactly equal to 1 milliliter, so densities are also often expressed in grams per milliliter (g/mL).

- ✔ **Pressure (an example is of gaseous reactants):** Pressure units are derived using the formula, Pressure = Force / Area. The SI units for force and area are Newtons (N) and square meters (m^2), so the SI unit of pressure, the Pascal (Pa), can be expressed as $N\,m^{-2}$.

Q. A physicist measures the density of a substance to be 20 kg/m^3. His chemist colleague, appalled with the excessively large units, decides to change the units of the measurement to the more familiar g/cm^3. What is the new expression of the density?

A. **0.02 g/cm^3.** A kilogram contains 1,000 grams, so 20 kilograms equals 20,000 grams. Well, 100 cm = 1m, therefore $(100 \text{ cm})^3 = (1\text{m})^3$. In other words, there are 100^3 (or 10^6) cubic centimeters in 1 cubic meter. Doing the division gives you 0.02 g/cm^3.

3. The Pascal, a unit of pressure, is equivalent to 1 Newton per square meter. If the Newton, a unit of force, is equal to 1 kilogram meter per second squared, what's the Pascal, expressed entirely in basic units?

Solve It

4. A student measures the length, width, and height of a sample to be 10 mm, 15 mm, and 5 mm respectively. If the sample has a mass of 0.9 Dg, what is the sample's density in g/mL?

Solve It

Converting between Units: The Conversion Factor

So what happens when chemist Reginald Q. Geekmajor neglects his SI units and measures the boiling point of his sample to be 101 degrees Fahrenheit, or the volume of his beaker to be 2 cups? Although Dr. Geekmajor should surely have known better, he can still save himself the embarrassment of reporting such dirty, unscientific numbers to his colleagues: He can use *conversion factors*.

A conversion factor simply uses your knowledge of the relationships between units to convert from one unit to another. For example, if you know that there are 2.54 centimeters in every inch (or 2.2 pounds in every kilogram, or 101.3 kilopascals in every atmosphere), then converting between those units becomes simple algebra. Peruse Table 2-3 for some useful conversion factors. And remember: If you know the relationship between any two units, you can build your own conversion factor to move between those units.

Table 2-3	Conversion Factors	
Unit	*Equivalent to*	*Conversion Factors*
Lengths		
Meter	3.3 feet	$\frac{3.3\,\text{ft}}{1\text{m}}$ or $\frac{1\text{m}}{3.3\,\text{ft}}$

Unit	Equivalent to	Conversion Factors
Lengths		
Foot	12 inches	$\frac{12\,in}{1\,ft}$ or $\frac{1\,ft}{12\,in}$
Inch	2.54 cm	$\frac{1\,in}{2.54\,cm}$ or $\frac{2.54\,cm}{1\,in}$
Volumes		
Gallon	16 cups	$\frac{16c}{1\,gal}$ or $\frac{1\,gal}{16c}$
Cup	237 mL	$\frac{237\,mL}{1c}$ or $\frac{1c}{237\,mL}$
Milliliter	1 cm^3	$\frac{1\,cm^3}{1\,mL}$ or $\frac{1\,mL}{1\,cm^3}$
Mass		
Kilogram	2.2 pounds	$\frac{2.2\,lb}{1\,kg}$ or $\frac{1\,kg}{2.2\,lb}$
Time		
Hour	3,600 seconds	$\frac{3600\,sec}{1\,hr}$ or $\frac{1\,hr}{3600\,sec}$
Pressure		
Atmosphere	101.3 kPa	$\frac{101.3\,kPa}{1\,atm}$ or $\frac{1\,atm}{101.3\,kPa}$
Atmosphere	760 mm Hg*	$\frac{760\,mm\,Hg}{1\,atm}$ or $\frac{1\,atm}{760\,mm\,Hg}$

*One of the more peculiar units you'll encounter in your study of chemistry is mm Hg, or millimeters of mercury, a unit of pressure. Unlike SI units, mm Hg don't fit neatly into the base-10 metric system, but reflect the way in which certain devices like blood pressure cuffs and barometers use mercury to measure pressure.

As with many things in life, chemistry isn't always as easy as it seems. Chemistry teachers are sneaky: They often give you quantities in non-SI units and expect you to use one or more conversion factors to change them to SI units — all this before you even attempt the "hard part" of the problem! We are at least marginally less sneaky than your typical chemistry teacher, but we hope to prepare you for such deception. So, expect to use this section throughout the rest of this book!

The following example shows how to use a basic conversion factor to "fix" non-SI units.

Q. Dr. Geekmajor absent-mindedly measures the mass of a sample to be 0.75 pounds and records his measurement in his lab notebook. His astute lab assistant wants to save the doctor some embarrassment, and knows that there are 2.2 pounds in every kilogram; the assistant quickly converts the doctor's measurement to SI units. What does she get?

A: **0.34 kg.**

$$\frac{0.75 \text{ lbs}}{1} \times \frac{1 \text{ kg}}{0.75 \text{ lbs}} = 0.34 \text{ kg}$$

Notice that something very convenient just happened. Because of the way this calculation was set up, you end up with pounds on both the top and bottom of the fraction. In algebra, whenever you find the same quantity in a numerator and in a denominator, you can cancel them out. Canceling out the pounds is a lovely bit of algebra, because you didn't want them around anyway. The whole point of the conversion factor is to get rid of an undesirable unit, transforming it into a desirable one — without breaking any rules. You had two choices of conversion factors to convert between pounds and kilograms; one with pounds on the top, and another with pounds on the bottom. The one to choose was the one with pounds on the bottom, so the undesirable pounds units cancel. Had you chosen the other conversion factor you would've ended up with

$$\frac{0.75 \text{ lbs}}{1} \times \frac{2.2 \text{ lbs}}{1 \text{ kg}} = 1.65 \text{lb}^2/\text{kg}$$

This calculation doesn't simplify your life at all, so it's clearly the wrong choice. If you end up with more complicated units after employing a conversion factor, then try the calculation again, this time flipping the conversion factor.

If you're a chemistry student, you're probably pretty familiar with the basic rules of algebra (nod your head in emphatic agreement . . . good). So, you know that you can't simply multiply one number by another and pretend that nothing happened — you altered the original quantity when you multiplied, didn't you? What in blazes is going on here? With all these conversion factors being multiplied willy-nilly, why aren't the International Algebra Police kicking down doors to chemistry labs all around the world?

Though a few would like you to believe otherwise, chemists can't perform magic. Recall another algebra rule: You can multiply any quantity by 1 and you'll always get back the original quantity. Now, look closely at the conversion factors in the example: 2.2 pounds and 1 kilogram are exactly the same thing! Multiplying by 2.2 lbs / 1 kg or 1 kg / 2.2 lbs is really no different than multiplying by 1.

Q. A chemistry student, daydreaming during lab, suddenly looks down to find that he's measured the volume of his sample to be 1.5 cubic *inches*. What does he get when he converts this quantity to cubic centimeters?

A. **24.6 cm^3.**

$$\frac{1.5 \text{ in}^3}{1} \times \left(\frac{2.54 \text{ cm}}{1 \text{ in}}\right)^3 = 24.6 \text{ cm}^3$$

You've already converted between inches and centimeters, but not between cubic inches and cubic centimeters. Rookie chemists often mistakenly assume that if there are 2.54 centimeters in every inch, then there are 2.54 cubic

centimeters in every cubic inch. No! Although this assumption seems logical at first glance, it leads to catastrophically wrong answers. Remember that cubic units are units of volume, and that the formula for volume is length × width × height. Imagine 1 cubic inch as a cube with 1-inch sides. The cube's volume is 1 in × 1 in × 1 in = 1 in^3.

Now consider the dimensions of the cube in centimeters: 2.54 cm x 2.54 cm x 2.54 cm. Calculate the volume using these measurements and you get 2.54 cm × 2.54 cm × 2.54 cm = 16.39 cm^3. This volume is much greater than 2.54 cm^3! Square or cube everything in your conversion factor, not just the units, and everything works out just fine.

5. A sprinter running the 100-meter dash runs how many feet?

6. At the top of Mount Everest, the air pressure is approximately 0.33 atmospheres, or one third of the air pressure at sea level. A barometer placed at the peak would read how many millimeters of mercury?

Solve It

7. A "league" is an obsolete unit of distance used by archaic (or nostalgic) sailors. A league is equivalent to 5.6 kilometers. If the characters in Jules Verne's novel *20,000 Leagues Under the Sea* actually travel to a depth of 20,000 leagues, how many kilometers under water are they? If the radius of the earth is 6,378 km, is this a reasonable depth? Why or why not?

Solve It

8. The slab of butter slathered by Paul Bunyan onto his morning pancakes is 2 feet wide, 2 feet long, and 1 foot thick. How many cubic meters of butter does Paul consume each morning?

Solve It

Letting the Units Guide You

In the earlier sections in this chapter, the problems have used just one conversion factor at a time. You may have noticed that Table 2-3 doesn't list all possible conversions (between meters and inches, for example). Rather than bother with memorizing or looking up conversion factors between every type of unit, you can memorize just a handful and use them one after another, letting the units guide you each step of the way.

Say you wanted to know the number of seconds in one year (clearly a very large number, so don't forget about your scientific notation). Very few people have this conversion memorized — or will admit to it — but everyone knows that there are 60 seconds in a minute, 60 minutes in an hour, 24 hours in a day, and 365 days in a year. So, use what you know to get what you want!

$$\frac{1 \text{ yr}}{1} \times \frac{365 \text{ days}}{1 \text{ yr}} \times \frac{24 \text{ hr}}{1 \text{ day}} \times \frac{60 \text{ min}}{1 \text{ hr}} \times \frac{60 \text{ sec}}{1 \text{ min}} = 3.15 \times 10^7 \text{ sec}$$

You can use as many conversion factors as you need as long as you keep track of your units in each step. The easiest way to do this is to cancel as you go, and use the remaining unit as a guide for the next conversion factor. For example, examine the first two factors of the years-to-seconds conversion. The years on the top and bottom cancel, leaving you with days. Because days remain on top, the next conversion factor needs to have days on the bottom and hours on the top. Canceling days then leaves you with hours, so your next conversion factor must have hours on the bottom and minutes on the top. Just repeat this process until you arrive at the units you want. Then do all the multiplying and dividing of the numbers and rest assured that the resulting calculation is the right one for the final units.

Q. A silly chemistry student measures the mass of a sample of mystery metal to be 0.5 pounds and measures the sample's dimensions to be 1 cubic inch. When she realizes her error, she attempts to convert her measurements into proper SI units. What should her units be, and what's the density of her sample (in those units)?

A. **The units should be g/mL or g/cm³, and the density is 13.9 g/mL.**

The proper SI units for density are g/mL (g/cm³), so she should use the following method:

$$\frac{0.5 \text{ lbs}}{1 \text{ in}^3} \times \left(\frac{1 \text{ in}}{2.54 \text{ cm}}\right)^3 \times \frac{1 \text{ cm}^3}{1 \text{ mL}} \times \frac{1 \text{ kg}}{2.2 \text{ lbs}} \times \frac{1000 \text{g}}{1 \text{ kg}} = 13.9 \text{ g/mL}$$

9. How many meters are in 15 feet?

Solve It

10. If Steve weighs 175 pounds, what's his weight in grams?

Solve It

11. How many liters are in 3 gallons of water?

12. If the dimensions of a cube of sample are 3 inches x 6 inches x 1 foot, what's the volume of that cube in cubic centimeters? Give your answer in scientific notation or use a metric system prefix.

Solve It

13. If there are 5.65 kilograms per every half liter of a particular substance, which of the following is that substance: liquid mercury (density 13.5 g/cm^3), lead (11.3 g/cm^3), or tin (7.3 g/cm^3)?

Solve It

Answers to Questions on Using and Converting Units

The following are the answers to the practice problems presented in this chapter.

1 1×10^7 **nm.** Both 10^2 centimeters and 10^9 nanometers equal 1 meter. Set the two equal to one another (10^2 cm = 10^9 nm) and solve for centimeters by dividing. This conversion tells you that 1 cm = $10^9 / 10^2$ nm, or 1×10^7 nm.

2 **2.5 kg.** Because there are 1,000 grams in 1 kilogram, simply divide 2,500 by 1,000 to get 2.5.

3 $1\,Pa = \dfrac{1\,kg}{ms^2}$

First, you write out the equivalents of Pascals and Newtons as explained in the problem:

$$1\,Pa = \frac{1N}{m^2} \text{ and } 1N = \frac{1\,kgm}{s^2}$$

Now, substitute Newtons (expressed in fundamental units) into the equation for the Pascal to

get $\quad 1\,Pa = \dfrac{\dfrac{1\,kgm}{s^2}}{m^2}$

Simplify this equation to $\quad 1\,Pa = \dfrac{1\,kgm}{m^2 s^2}\quad$ and cancel out the meter, which appears in both the top

and the bottom, leaving $\quad Pa = \dfrac{1\,kg}{ms^2}$

4 **12 g/mL.** Because a milliliter is equivalent to 1 cubic centimeter, the first thing to do is to convert all the length measurements to centimeters: 1 cm, 1.5 cm, and 0.5 cm. Then, multiply the converted lengths to get the volume: 1 cm $\times$ 1.5 cm $\times$ 0.5 cm = 0.75 cm^3, or 0.75 mL. The mass should be expressed in grams rather than Dg; there are 10 grams in 1 decogram, so 0.9 Dg = 9g. Using the formula D = m / V, you calculate a density of 9 grams per 0.75 milliliter, or 12 g/mL.

5 **330m.** Set up the conversion factor as shown.

$$\frac{100m}{1} \times \frac{3.3\,ft}{1m} = 330ft$$

6 **251 mm Hg.**

$$\frac{0.33\,atm}{1} \times \frac{760\,mm\,Hg}{1\,atm} = 251mm\,Hg$$

7 1.12×10^5 **km.**

$$\frac{20{,}000 \text{ leagues}}{1} \times \frac{5.6 \text{ km}}{1 \text{ league}} = 1.12 \times 10^5 \text{km}$$

The radius of the Earth is only 6,378 km. 20,000 leagues is 17.5 times that radius! So, the ship would've burrowed through the Earth and been halfway to the orbit of Mars if it had truly sunk to such a depth. Jules Verne intended the title to imply a distance traveled across the sea, not a depth!

8   **0.11 m^3.** The volume of the butter in feet is 2 ft $\times$ 2 ft $\times$ 1 ft, or 4 ft^3.

$$\frac{4 \text{ ft}^3}{1} \times \left(\frac{1m}{3.3 \text{ ft}} \right)^3 = 0.11 m^3$$

9   **4.572m.**

$$\frac{15 \text{ ft}}{1} \times \frac{12 \text{ in}}{1 \text{ ft}} \times \frac{2.54 \text{ cm}}{1 \text{ in}} \times \frac{1m}{100 \text{ cm}} = 4.572m$$

10 7.95×10^4**g.**

$$\frac{175 \text{ lbs}}{1} \times \frac{1 \text{ kg}}{2.2 \text{ lbs}} \times \frac{1000g}{1 \text{ kg}} = 7.95 \times 10^4 \text{g}$$

11   **11.4L.**

$$\frac{3 \text{ gal}}{1} \times \frac{16c}{1 \text{ gal}} \times \frac{237 \text{ mL}}{1 \text{ cup}} \times \frac{1L}{1000 \text{ mL}} = 11.4L$$

12 3.54×10^3 **cm^3.** First convert all of the inch and foot measurements to centimeters:

$$\frac{3 \text{ in}}{1} \times \frac{2.54 \text{ cm}}{1 \text{ in}} = 7.62 \text{ cm} \text{ by } \frac{6 \text{ in}}{1} \times \frac{2.54 \text{ cm}}{1 \text{ in}} = 15.24 \text{ cm} \text{ by } \frac{1 \text{ ft}}{1} \times \frac{12 \text{ in}}{1 \text{ ft}} \times \frac{2.54 \text{ cm}}{1 \text{ in}} = 30.48 \text{ cm}$$

The volume is therefore 7.62 cm $\times$ 15.24 cm $\times$ 30.48 cm, or 3.54×10^3 cm^3.

13   **The substance is lead.**

$$\frac{5.65 \text{ kg}}{0.5 \text{ L}} \times \frac{1 \text{ L}}{1000 \text{ mL}} \times \frac{1 \text{ mL}}{1 \text{ cm}^3} \times \frac{1000 \text{ g}}{1 \text{ kg}} = 11.3 \text{ g/cm}^3 \text{, which is exactly the density of lead.}$$

Chapter 3

Organizing Matter into Atoms and Phases

....

In This Chapter

▶ Peeking inside the atom: Protons, electrons, and neutrons

▶ Deciphering atomic numbers and mass numbers

▶ Understanding isotopes and calculating atomic masses

▶ Ordering matter within different phases

....

"*B*ig stuff is built from smaller pieces of stuff. If you keep breaking stuff down into smaller and smaller pieces, eventually you'll reach the smallest possible bit of stuff. Let's call that bit an atom." This is how the Greek philosopher Democritus might have explained his budding concept of "atomism" to a buddy over a flask of Cretan wine. Like wine, the idea had legs.

For hundreds of years, scientists have operated under the idea that all matter is made up of smaller building blocks called *atoms*. So small, in fact, that until the invention of the electron microscope in 1931, the only way to find out anything about these tiny, mysterious particles was to design a very, very clever experiment. Chemists couldn't exactly corner a single atom in a back alley somewhere and study it alone — they had to study the properties of whole gangs of atoms and try to guess what individual ones might be like. Through remarkable ingenuity and incredible luck, chemists now understand a great deal about the atom. After reading this chapter, so will you.

You also discover that matter keeps you on your toes by shape-shifting. Samples of matter can change from sturdy solids to loose liquids to ghostly gases, depending on conditions. All three of these phases of matter have very specific comfort zones where they're most likely to exist, and by the end of this chapter, you'll be able to predict likely phases based on temperature and pressure.

Building Atoms from Subatomic Particles

For now, picture an atom as a microscopic Lego. Atoms come in a variety of shapes and sizes, and you can build larger structures out of them. Like a Lego, an atom is extremely hard to break. In fact, so much energy is stored inside atoms that breaking them in half results in a nuclear explosion. Boom!

An atom is made of smaller pieces, called *subatomic particles,* but it's still considered the smallest possible unit of an element, because after you break an atom of an element into subatomic particles, the pieces lose the unique properties of that element.

Virtually all substances are made of atoms. The universe seems to use about 120 unique atomic Lego blocks to build neat things like galaxies and people and whatnot. All atoms are made of the same three subatomic particles: the *proton*, the *electron,* and the *neutron*. Different types of atoms (in other words, different *elements*) have different combinations of these particles, which gives each element unique properties. For example:

✔ **Atoms of different elements have different masses.** Atomic masses are measured in multiples of the mass of a single proton, called *atomic mass units* (equivalent to 1.66×10^{-27} kg), or *amu*. (We discuss atomic mass in more detail in the later section "Accounting for Isotopes Using Atomic Masses.")

✔ **Two of the subatomic particles — the proton and the electron — have a charge, so atoms with different numbers of these two particles have different atomic charges.** Atomic charges are measured in multiples of the charge of a single proton.

The must-know information about the three subatomic particles is summarized in Table 3-1.

Table 3-1	The Subatomic Particles	
Particle	*Mass*	*Charge*
Proton	1 amu	+1
Electron	$\frac{1}{1836}$ amu	−1
Neutron	1 amu	0

Notice in Table 3-1 that protons and electrons have equal and opposite charges, and that neutrons are neutral. Atoms always have an equal number of protons and electrons, so the overall charge of an atom is neutral (that is to say, zero). Many atoms actually prefer to gain or lose electrons, which causes them to gain a nonzero charge; in other words, the number of negative charges is no longer balanced by the number of positive charges. Charged atoms are called *ions* and are explained in Chapter 5. Until then, assume that all of our atoms have equal numbers of protons and electrons.

Now look at Table 3-1 with an eye to mass. Protons and neutrons have the same mass. Electrons have nearly 2,000 times less mass. This means that most of an atom's mass comes from protons and neutrons. Although electrons contribute a lot of negative charge, they contribute very little mass.

Fine, you say. It's all well and good that the subatomic particles have all these lovely properties, but what does an atom actually look like? Generations of ingenious scientists have tackled this question. The result of all the clever experimentation and tricky math has been a series of models, each a bit more refined than the one before. The models are milestones in a scientific story. Pop some corn, and read on.

J. J. Thomson: Cooking up the "plum pudding" model

The first subatomic particle was discovered by J. J. Thomson in the late 1800s. Thomson performed a series of experiments using a device called a cathode ray tube, a contraption that eventually evolved into the modern television. Thomson realized that a beam of electrons (or a cathode ray, in scientific speak) could be deflected in a magnetic field. This result and others got Thomson thinking that the whole "indivisible atom" model had its limitations; atoms were actually composed of other, subatomic particles. Thomson proposed a model of the atom, called the "plum pudding" model. You've probably never had the misfortune to taste plum pudding, a traditional English dessert consisting of dried plums mixed into a thick pudding. Thomson's model was as important to chemistry as the dessert is foul tasting.

Thomson, like all chemists of his day, knew that two negative charges repel one another, so he pictured the atom as a collection of evenly spaced, negatively charged particles. Thomson called these charges *corpuscles,* but Thomson's corpuscles have subsequently been awarded the much nicer name *electrons.* Thomson also knew that the atom was electrically neutral (meaning it has zero overall charge), so he figured the atom must also contain an amount of positive charge equal to the negative charge of the electrons. Because the proton had not yet been discovered, Thomson imagined this positive charge as a soup in which all of the negative electrons were suspended, just like plums are suspended in plum pudding.

Ernest Rutherford: Shooting at gold

The next leap forward was made by Ernest Rutherford in 1909. Rutherford set out to test Thomson's plum pudding model of the atom. To do so, Rutherford made an extremely thin sheet of gold foil and shot alpha particles (helium nuclei) at it. This may sound a little crazy, but there was method to Rutherford's madness. If the plum pudding model were true, Rutherford expected that when alpha particles crashed into the gold foil, loosely bound electrons of the gold atoms would deflect the alpha particles by a few degrees at most. When Rutherford attempted it, however, he found that 1 in every 8,000 (or so) alpha particles deflected by 90 degrees or more! He famously compared the result to watching a shot bullet bounce off a piece of tissue paper, a result perplexing enough to drop the jaw of even a jaded chemist.

After much pondering, Rutherford eventually realized that most of an atom must be empty space, and that most of the atom's mass — and one of its two kinds of charge — must be concentrated at the center. So 7,999 of every 8,000 particles Rutherford shot at the foil missed this tiny bundle of mass, passing straight through, but every so often a lucky shot smashed into the nucleus and was deflected at a large angle.

We now know that the positive charge of an atom is concentrated at its center in the form of protons. The protons reside there along with all of the atom's neutrons. There is something very strange and counterintuitive about this idea. Remember that like charges repel one another; so having all the positive charge of an atom concentrated in one, tiny, central area is truly bizarre. A very, very strong force must be holding together all those positive charges. This force, unimaginatively dubbed the *strong force,* is something we can take for granted; nuclear physicists can't take it for granted, but that's their headache.

Niels Bohr: Comparing the atom to the solar system

Enter Niels Bohr in the early 1900s. Dr. Bohr was not an experimentalist like Thomson or Rutherford. Bohr was a *theorist,* which basically means he sat around pondering things. Having pondered his way to an "aha" moment, his job became to prove his ideas mathematically.

Bohr was aware of Rutherford's gold foil experiment. It occurred to Bohr that the atom may operate very much like the solar system, with most of the mass concentrated in the center (at the sun), with smaller bodies (the planets) orbiting the center at specific distances. According to this model, low-mass electrons in an atom orbit the central nucleus, which contains all the massive protons and neutrons. This conceptual leap was neat, logical, and crazy. The understanding of electricity and magnetism at the time suggested that there was no way for a negative charge to orbit at a constant distance from a positive charge. Classic theories suggested that an orbiting electron would eventually spiral into a central nucleus.

Bohr bypassed this problem with a neat little trick called "quantization of angular momentum." Basically, Bohr invented a whole new set of rules for how electrons should behave in an atom. Strangely enough, the predictions of his mathematical model matched experiments so well that nobody could prove him wrong. Although now an entire branch of physics, called *quantum mechanics,* is much more accurate than Bohr's model, his predictions are so nearly right and so convenient that we gratefully leave quantum mechanics to math-happy physicists. We'll stick with the picture painted for us by Bohr.

Q. Rutherford's gold atoms contained 197 nuclear particles, 79 of which were protons. How many neutrons and how many electrons did each gold atom have?

A. **118 neutrons and 79 electrons.** The nucleus contains all of the protons and neutrons in an atom, so if 79 of the 197 particles in a gold nucleus are protons, the remaining 118 particles must be neutrons.

All atoms are electrically neutral, so there must be a total of 79 electrons (in other words, 79 negative charges) to balance out the 79 positive charges of the protons. This type of logic leads us to a general formula that can be used to calculate proton or neutron counts. This formula is $M = P + N$, where M is the atomic mass, P is the number of protons, and N is the number of neutrons.

---

1. If an atom has 71 protons, 71 electrons, and 104 neutrons, how many particles reside in the nucleus and how many outside of the nucleus?

Solve It

2. If an atom's nucleus weighs 31 amu and contains 15 protons, how many neutrons and electrons does the atom have?

Solve It

3. What is the mass in kilograms of the uranium nucleus (the heaviest of all naturally occurring atoms) if it contains 92 protons and 146 neutrons?

Solve It

Deciphering Chemical Symbols: Atomic and Mass Numbers

Two very important numbers tell us much of what we need to know about an atom: the *atomic number* and the *mass number*. Chemists tend to memorize these numbers like baseball fans memorize batting averages, but clever chemistry students like you need not resort to memorization. You have the ever-important periodic table of the elements at your disposal. We discuss the logical structure and organization of the periodic table in detail in Chapter 4, so for now we simply introduce the meaning of the atomic and mass numbers without going into great detail about their consequences.

Atomic numbers are like name tags — they identify an element as carbon, nitrogen, beryllium, and so on by telling you the number of protons in the nucleus of that element. By the numbers of their protons are atoms known. Adding a proton or removing one from the nucleus of an atom changes the elemental identity of an atom.

Atoms are very fond of their identities, so simply adding or subtracting protons is quite difficult. Certain heavy elements can be split, and certain light elements can be smashed together, in processes called *nuclear fission* and *nuclear fusion,* respectively. Splitting or joining atomic nuclei releases a tremendous amount of energy and isn't something you should try at home, even if you *do* have a nuclear reactor in your basement.

The only other way for an atom to change its atomic number (and therefore its identity) is to *decay* through a radioactive process. For most nuclei, this decay process happens about as often as a chemist enters a high-end hair salon. Not that there's anything wrong with that.

In the periodic table, you find the atomic number above the one- or two-letter abbreviation for an element. The abbreviation is the element's chemical symbol. Notice that the elements of the periodic table are lined up in order of atomic number, as if they've responded to some sort of roll call. Atomic number increases by 1 each time you move to the right in the periodic table; when a row ends, the sequence of increasing atomic numbers begins again at the left side of the next row down. You can check out the periodic table for yourself in Chapter 4.

The second identifying number of an atom is its mass number. The mass number reports the mass of the atom's nucleus in amu. Because protons and neutrons weigh 1 amu each (as you find out earlier in this chapter), the mass number equals the sum of the numbers of protons and neutrons.

Why, you may wonder, don't we care about the mass of the electrons? Is some sort of insidious subatomic particle-ism afoot? No. An electron has only $\frac{1}{1,836}$ the mass of a proton or neutron. So, to make mass numbers nice and even, chemists have decided to conveniently forget that electrons have mass. Although this assumption is not, well, true, the contributions of electrons to the mass of an atom are so small that the assumption is usually harmless.

To specify the atomic and mass numbers of an element, chemists typically write the symbol of the element in the form $_Z^A X$, where Z is the atomic number, A is the mass number, and X is the chemical symbol for that element.

Q. What is the name, atomic number, mass number, number of protons, number of electrons, and number of neutrons of each of the following four elements: $_{17}^{35}Cl$, $_{17}^{37}Cl$, $_{76}^{190}Os$, and $_{19}^{39}K$?

A. The answers to questions like these, favorites of chemistry teachers, are best organized in a table. First, look up the symbols Cl, Os, and K in the periodic table (see Figure 3–1) and find the names of these elements. Enter what you find in the first column. To fill in the second and third columns, read the atomic number and mass number from the lower left and upper left of the chemical symbols given in the question. The atomic number equals the number of protons; the number of electrons is the same as the number of protons, because elements have zero overall charge. So, fill in the fourth and fifth columns with the same numbers you entered in column two. Lastly, subtract the atomic number from the mass number to get the number of neutrons, and enter that value into column six. Voilá! The entire private lives of each of these atoms is now laid before you. Your answer should look like the following table.

Name	Atomic Number	Mass Number	Number of Protons	Number of Electrons	Number of Neutrons
Chlorine	17	35	17	17	18
Chlorine	17	37	17	17	19
Osmium	76	190	76	76	114
Potassium	19	39	19	19	20

4. Write the proper chemical symbol for an atom of bismuth with a mass of 209 amu.

Solve It

5. Fill in the following chart for $^{1}_{1}Cl$, $^{32}_{24}Cr$, $^{192}_{77}Ir$, and $^{96}_{42}Mo$.

Name	Atomic Number	Mass Number	Number of Protons	Number of Electrons	Number of Neutrons

6. Write the $^{A}_{Z}X$ form of the two elements in the following table:

Name	Atomic Number	Mass Number	Number of Protons	Number of Electrons	Number of Neutrons
Tungsten	74	184	74	74	110
Lead	82	207	82	82	125

Solve It

7. Use the periodic table and your knowledge of atomic numbers and mass numbers to fill in the missing pieces in the following table:

Name	Atomic Number	Mass Number	Number of Protons	Number of Electrons	Number of Neutrons
Silver		108			
	16				16
		64	29		
				18	22

Accounting for Isotopes Using Atomic Masses

It's Saturday night. The air is charged with possibility. Living in the moment, you peruse your personal copy of the periodic table. What's this? You notice the numbers that appear below the atomic symbols appear to be related to the elements' mass numbers — but they're not nice whole numbers. What could it all mean?

In the previous section, we explain that the mass number of an atom equals the sum of the numbers of protons and neutrons in the atom's nucleus. So, how can you have a fractional number of protons or neutrons? There's no such thing as half a proton, nor 0.25 neutron. Your chemist's sense of meticulousness has been offended, and you demand answers.

As it turns out, most elements come in several different varieties, called *isotopes*. Isotopes are atoms of the same element that have different mass numbers; the differences in mass number arise from different numbers of neutrons. The messy looking numbers with all those decimal places are *atomic masses*. An atomic mass is a weighted average of the masses of all the naturally occurring isotopes of an element. Chemists have measured the percentage of each element that exists in different isotopic forms. In the weighted average of the atomic mass, the mass of each isotope contributes in proportion to how often that isotope occurs in nature. More common isotopes contribute more to the atomic mass.

But there's more to the story of isotopes. Chemists are titillated by them. For Bohr's sake, why? A neutron is, after all, a *neutral* particle, so you wouldn't think adding or subtracting one would change much about an atom. Indeed, much of the time, adding a neutron doesn't significantly alter the properties of an element. But occasionally, adding but one seemingly innocent little neutron can have sinister effects, pushing an atom over to the dark side, into the realm of *radioactivity*. If an atom has just the right (or, perhaps, *wrong*) number of neutrons, it becomes unstable. When it comes to atomic nuclei, instability means radioactivity. Unstable nuclei break down (or decay) into more stable forms. The less stable the nucleus, the faster it tends to decay. When unstable nuclei decay, they tend to emit particles and energy. These emissions are the source of radioactivity. The word *radioactivity* conjures up images of three-headed frogs and fish with ten eyes, but radioactivity doesn't deserve its bad reputation. Many radioactive elements are harmless, and many have useful properties.

Consider the element carbon, for example. Carbon occurs naturally in three isotopes:

- **Carbon-12** ($_{6}^{12}C$, or carbon with six protons and six neutrons) is boring, old, run-of-the-mill carbon, accounting for 99 percent of all the carbon out there.

- **Carbon-13** ($_{6}^{13}C$, or carbon with six protons and *seven* neutrons) is a slightly more rare (though still dull) isotope, accounting for most of the remaining 1 percent of carbon atoms. Taking on an extra neutron makes carbon-13 slightly heavier than carbon-12, but does little else to change its properties.

  However, even this minor change has some useful consequences. Scientists compare the ratio of carbon-12 to carbon-13 within meteorites to help determine their origin.

- **Carbon-14** ($_{6}^{14}C$, or carbon with six protons and *eight* neutrons) shows its interesting little face in only one out of every trillion or so carbon atoms. So if you're thinking you won't be working with large samples of carbon-14 in your chemistry lab, you're right!

Carbon-14, the most exotic and interesting isotope of carbon, is important in a technique called *radioactive dating*. Carbon is the critical building block of organic molecules, including those found in animal bodies. Because bodies contain trillions of trillions of trillions of carbon atoms, the total amount of the rare carbon-14 is detectable. What's more, carbon-14 is a *radioisotope,* meaning that as it decays, it emits radiation. Before you go checking the mirror to search for traces of a third eye, rest assured that the levels of naturally occurring carbon-14 within your body won't harm you. But those levels are sufficient to be useful to scientists who want to determine the age of fossils. How?

Imagine Serena, a sultry Stone Age woman who met her untimely end 50,000 years ago, caught tragically in the midst of a violent dispute between two of her suitors. While Serena was alive, the carbon in her body was continually replenished, so that she consisted of about 99 percent carbon-12, 1 percent carbon-13, and 0.0000000001 percent carbon-14. When Serena kicked the bucket, so did the biological processes that replenished the carbon in her body. With their supply of carbon-14 cut off, Serena's bones slowly became depleted of carbon-14, as that isotope decayed into nitrogen.

Millennia later, paleontologists uncovering the dramatic site of Serena's final moments can dash straight to their friendly neighborhood chemist, Dr. Isotopian, with samples of Serena in hand. With pride, and hiking up his pants to an unfashionable degree, Dr. Isotopian is able to report Serena's approximate age. Radioisotopes decay at highly predictable rates. By measuring the relative levels of carbon isotopes in the sample of Serena, Dr. Isotopian can calculate the time passed since Serena perished. Different elements have isotopes that decay at different rates, so different isotopes can be used to date samples of wildly different ages.

REMEMBER

Precise measurements of the amounts of different isotopes can be important. You need to know the exact measurements if you're asked to figure out an element's atomic mass. To calculate an atomic mass, you need to know the masses of the isotopes and the percent of the element that occurs as each isotope (this is called the *relative abundance*). To calculate an average atomic mass, make a list of each isotope along with its mass and its percent relative abundance. Multiply the mass of each isotope by its relative abundance. Add the products. The resulting sum is the atomic mass.

TIP

Certain elements, such as chlorine, occur in several very common isotopes, so their average atomic mass isn't close to a whole number. Other elements, such as carbon, occur in one very common isotope and several very rare ones, resulting in an average atomic mass that is very close to the whole-number mass of the most common isotope.

EXAMPLE

Q. Chlorine occurs in two common isotopes. It appears as $^{35}_{17}\text{Cl}$ 75.8% of the time and as $^{37}_{17}\text{Cl}$ 24.2% of the time. What is its average atomic mass?

A. **35.48 amu.** First, multiply each atomic mass by its relative abundance, using the fractional form (75.8% = 0.758):

35 amu × 0.758 = 26.53 amu

37 amu × 0.242 = 8.95 amu

Then, add the two results together to get the average atomic mass. Compare the value to the value on your periodic table. If you have done the calculation correctly, the two values should match.

26.53 amu + 8.95 amu = 35.48 amu

8. Hydrogen occurs naturally in three different isotopes, $_1^1H$, $_1^2H$ (commonly called deuterium), and $_1^3H$ (commonly called tritium). Look up the average atomic mass of hydrogen on your periodic table. What does this mass tell you about the relative abundances of these isotopes? Which isotope is the most common?

Solve It

9. Magnesium occurs in three fairly common isotopes: $_{12}^{24}Mg$, $_{12}^{25}Mg$, and $_{12}^{26}Mg$, which have percent abundances of 78.9%, 10.0%, and 11.1% respectively. Calculate the average atomic mass of magnesium.

Solve It

Moving between the Phases of Solids, Liquids, and Gases

 Elements are made up of atoms, and elements can exist in one of three common forms or *phases:* solid, liquid, or gas. At room temperature, which is about 25 degrees Celsius (or 77 degrees Fahrenheit), and at atmospheric pressure, the majority of the elements on the periodic table are either gases or solids. Only two elements, bromine and mercury, exist as liquids under normal conditions. Look over Figure 3-1 to get a general feel for where on the periodic table these solids, liquids, and gases hang out. Notice in particular that hydrogen appears to have been kicked out of the gas clique that claims the right-hand side of the periodic table.

As temperature and pressure change, an element can be pushed over the boundary from one phase to another. Typically, low temperatures and high pressures promote solid forms, and high temperatures and low pressures promote gaseous forms. Liquids occupy the middle ground.

In chemistry, discussions of temperature often refer to the *Kelvin scale,* which you may not have previously encountered. Boldly, the Kelvin scale sets its zero at a very extreme condition, called *absolute zero.* There is indeed something very absolute about this zero, because it's the temperature at which all particles freeze, ceasing all movement. Absolute zero is the lowest possible temperature, representing the complete absence of heat. It's as low as you can possibly go. If you could reduce your temperature to absolute zero, you wouldn't age. On the other hand, you wouldn't be alive. Any takers?

G **H** Hydrogen																	G **He** Helium
S **Li** Lithium	S **Be** Beryllium											S **B** Boron	S **C** Carbon	G **N** Nitrogen	G **O** Oxygen	G **F** Fluorine	G **Ne** Neon
S **Na** Sodium	S **Mg** Magnesium											S **Al** Aluminum	S **Si** Silicon	S **P** Phosphorus	S **S** Sulfur	G **Cl** Chlorine	G **Ar** Argon
S **K** Potassium	S **Ca** Calcium	S **Sc** Scandium	S **Ti** Titanium	S **V** Vanadium	S **Cr** Chromium	S **Mn** Manganese	S **Fe** Iron	S **Co** Cobalt	S **Ni** Nickel	S **Cu** Copper	S **Zn** Zinc	S **Ga** Gallium	S **Ge** Germanium	S **As** Arsenic	S **Se** Selenium	L **Br** Bromine	G **Kr** Krypton
S **Rb** Rubidium	S **Sr** Strontium	S **Y** Yttrium	S **Zr** Zirconium	S **Nb** Niobium	S **Mo** Molybdenum	S **Tc** Technetium	S **Ru** Ruthenium	S **Rh** Rhodium	S **Pd** Palladium	S **Ag** Silver	S **Cd** Cadmium	S **In** Indium	S **Sn** Tin	S **Sb** Antimony	S **Te** Tellurium	S **I** Iodine	G **Xe** Xenon
S **Cs** Cesium	S **Ba** Barium	S **La** Lanthanum	S **Hf** Hafnium	S **Ta** Tantalum	S **W** Tungsten	S **Re** Rhenium	S **Os** Osmium	S **Ir** Iridium	S **Pt** Platinum	S **Au** Gold	L **Hg** Mercury	S **Tl** Thallium	S **Pb** Lead	S **Bi** Bismuth	S **Po** Polonium	S **At** Astatine	G **Rn** Radon
S **Fr** Francium	S **Ra** Radium	S **Ac** Actinium	S **Rf** Rutherfordium	S **Db** Dubnium	S **Sg** Seaborgium	S **Bh** Bohrium	S **Hs** Hassium	S **Mt** Meitnerium	S **Ds** Darmstadtium	S **Rg** Roentgenium	S **Uub** Ununbium	S **Uut** Ununtrium	S **Uuq** Ununquadium	S **Uup** Ununpentium	S **Uuh** Ununhexium	**Uus** Ununseptium (Not yet observed)	G **Uuo** Ununoctium

S **Ce** Cerium	S **Pr** Praseodymium	S **Nd** Neodymium	S **Pm** Promethium	S **Sm** Samarium	S **Eu** Europium	S **Gd** Gadolinium	S **Tb** Terbium	S **Dy** Dysprosium	S **Ho** Holmium	S **Er** Erbium	S **Tm** Thulium	S **Yb** Ytterbium	S **Lu** Lutetium
S **Th** Thorium	S **Pa** Protactinium	S **U** Uranium	S **Np** Neptunium	S **Pu** Plutonium	S **Am** Americium	S **Cm** Curium	S **Bk** Berkelium	S **Cf** Californium	S **Es** Einsteinium	S **Fm** Fermium	S **Md** Mendelevium	S **No** Nobelium	S **Lr** Lawrencium

KEY
S = **Solid** at room temperature
L = **Liquid** at room temperature
G = **Gas** at room temperature

Figure 3-1: The solids, liquids, and gases on the periodic table.

In contrast to the Kelvin scale, the *Celsius scale* sets its zero at the freezing point of water. Aside from that shift in zero-point, the two scales are identical. The size of a Celsius degree is exactly equal to the size of a Kelvin degree. A Celsius temperature equals a Kelvin temperature plus 273 degrees. The freezing point of water is therefore a balmy 273 Kelvin (K).

At absolute zero, all elements are solids. At all temperatures greater than about 6,000K, all elements are gaseous. Table 3-2 gives a sampling of common elements and the temperatures at which they are solids, liquids, and gases.

Table 3-2	A Sample of Elemental Phases		
Element	**Solid**	**Liquid**	**Gas**
Solids at Room Temperature (298K):			
Sodium	0–371K	371–1,156K	1,156K+
Gold	0–1,337K	1,337–3,129K	3,129K+
Iron	0–1,811K	1,811–3,134K	3,134K+
Liquids at Room Temperature (298K):			
Mercury	0–234K	234–630K	630K+
Bromine	0–267K	267–332K	332K+
Gases at Room Temperature (298K):			
Hydrogen	0–14K	14–21K	21K+
Oxygen	0–55K	55–90K	90K+
Nitrogen	0–63K	63–77K	77K+

Throughout this book, you often see a letter in parentheses following the chemical symbol of an element, such as C(*s*), Br(*l*), or Ne(*g*). These designations are fairly straightforward: (*s*) means the element is a solid, (*l*) means liquid, and (*g*) means gas.

We talk about transitions between phases at various temperatures and pressures in much more detail in Chapter 10.

Q. At what temperature in degrees Celsius does bromine change phase from a liquid to a gas?

A. **Bromine becomes a gas at 59°C.** To answer this question, refer to Table 3-2. The table tells us that bromine is a solid at temperatures below 267K, a liquid at temperatures between 267K and 332K, and a gas at temperatures above 332K. Therefore, bromine undergoes a phase transition from solid to liquid at 267K and another transition from liquid to gas at 332K. Because the liquid-gas transition is the one concerning our question, simply convert 332K into a Celsius temperature by subtracting 273.

10. If the element cesium is a solid at temperatures below 575°C, a liquid between 575°C and 1,217°C, and a gas above 1,217°C, at what temperature in Kelvin does each phase change occur?

Solve It

11. If the temperature of a sample of iron is 2,000°C, in what phase is the sample?

 Solve It

Answers to Questions on Organizing Matter

Now that you've organized the facts on atoms and phases into clearly labeled compartments in your brain, see how accessible that information is. Check your answers to the practice problems presented in this chapter.

1   **175 inside, 71 outside.** The nucleus of an atom consists of protons and neutrons, so this atom (lutetium) has 175 particles in its nucleus. Electrons are the only subatomic particles that aren't included in the nucleus, so lutetium has 71 particles outside of its nucleus.

2   **16 neutrons, 15 electrons.** A nuclear mass of 31 amu means that the nucleus has 31 particles. Because 15 of them are protons, that leaves 16 amu for the neutrons. The number of protons and electrons are equal in a neutral atom (in this case, we're talking about phosphorus), so the atom has 15 electrons.

3 3.95×10^{-25} **kg.** First, calculate the atom's mass number, which is 92 amu + 146 amu = 238 amu. Next, convert this mass to kilograms by using the conversion factor 1 amu = 1.66×10^{-27} kg. That's right! The unit conversion haunts you once more! Set up this problem just as described in Chapter 2.

$$\frac{238 \text{ amu}}{1} \times \frac{1.66 \times 10^{-27} \text{kg}}{1 \text{ amu}} = 3.95 \times 10^{-25} \text{kg}$$

4 $_{83}^{209}\text{Bi}.$ Find bismuth in the periodic table to get its chemical abbreviation and its atomic number. Because you already have its mass number, all you need to do is write all this information into $_{Z}^{A}\text{X}$ form.

5

Name	Atomic Number	Mass Number	Number of Protons	Number of Electrons	Number of Neutrons
Hydrogen	1	1	1	1	0
Chromium	24	52	24	24	28
Iridium	77	192	77	77	115
Molybdenum	42	96	42	42	54

6 $_{74}^{184}\text{W},\ _{82}^{207}\text{Pb}.$ Notice that you don't need most of the information in the table; all you really need to look up is the chemical symbol of each element.

7 When you're given the atomic number, number of protons, or number of electrons, you automatically know the other two numbers because they're all equal. Each element in the periodic table is listed with its atomic number, so by locating the element, you can simply read off the atomic number and therefore the number of protons and electrons. To calculate the atomic mass or the number of neutrons, you must be given one or the other. Calculate atomic mass by adding the number of protons to the number of neutrons. Alternately, calculate the number of neutrons by subtracting the number of protons from the atomic mass.

Name	*Atomic Number*	*Mass Number*	*Number of Protons*	*Number of Electrons*	*Number of Neutrons*
Silver	47	108	47	47	61
Sulfur	16	32	16	16	16
Copper	29	64	29	29	35
Argon	18	40	18	18	22

8 $^1_1$**H .** The average atomic mass of hydrogen (which you can look up in the periodic table) is 1.0079 amu. This mass is so close to 1 that you know that the most commonly occurring isotope of hydrogen, by far, must be the isotope with a mass number of 1. This isotope is $^1_1$H .

9 **24.33 amu.** First, multiply the three mass numbers by their relative abundances in fractional form.

24 amu × 0.789 = 18.94 amu

25 amu × 0.100 = 2.50 amu

26 amu × 0.111 = 2.89 amu

Then add the resulting products to get the average atomic mass.

10 **848K, 1490K.** The solid-liquid phase change for cesium occurs at 575°C, and the liquid-gas phase change occurs at 1,217°C. To convert these temperatures to Kelvin, simply subtract 273 from each.

11 **Liquid.** Here again, the first step is to convert Celsius to Kelvin by subtracting 273; the sample has a temperature of 1,727K. According to Table 3-2, the sample is therefore in the liquid phase.

Chapter 4

Surveying the Periodic Table of the Elements

In This Chapter

▶ Moving up, down, left, and right on the periodic table

▶ Grasping the value of valence electrons

▶ Taking stock of electron configurations

▶ Equating an electron's energy to light

There it hangs, looming ominously over the chemistry classroom, a formidable wall built of bricks with names like "C," "Ag," and "Tc." Behold, it's the *periodic table of the elements*. But don't be fooled by its stern appearance or intimidated by its teeming details. The table is your friend, your guide, your key to making sense of chemistry. To begin to make friends with the table, concentrate on its trends. Start simply: Notice that the table has rows and columns. Keep your eye on the columns and rows, and soon you'll be making sense of things like electron affinity and atomic radii. Really.

Reading Periods and Groups in the Periodic Table

Take a look at your new friend, the periodic table, shown in Figure 4-1. Notice the horizontal rows and the vertical columns of elements.

✔ The rows are called *periods*.

✔ The columns are called *groups*.

As you move across any period, you pass over a series of elements whose properties change in a predictable way — predictability is the table's way of showing you that it likes you. The elements within any group have very similar properties. The properties of the elements emerge mostly from their different numbers of protons and electrons (see Chapter 3 for a refresher), and from the arrangement of their electrons. We explain how to predict the properties of elements from looking at the periodic table in the next section; for now, we just want to describe how the elements are arranged in different periods and groups.

Figure 4-1: The periodic table of the elements.

PERIODIC TABLE OF THE ELEMENTS

	1 IA	2 IIA	3 IIIB	4 IVB	5 VB	6 VIB	7 VIIB	8 VIIIB	9 VIIIB	10 VIIIB	11 IB	12 IIB	13 IIIA	14 IVA	15 VA	16 VIA	17 VIIA	18 0
1	1 H Hydrogen 1.00797																	He Helium 4.0026
2	3 Li Lithium 6.939	4 Be Beryllium 9.0122											5 B Boron 10.811	6 C Carbon 12.01115	7 N Nitrogen 14.0067	8 O Oxygen 15.9994	9 F Fluorine 18.9984	10 Ne Neon 20.183
3	11 Na Sodium 22.9898	12 Mg Magnesium 24.312											13 Al Aluminum 26.9815	14 Si Silicon 28.086	15 P Phosphorus 30.9738	16 S Sulfur 32.064	17 Cl Chlorine 35.453	18 Ar Argon 39.948
4	19 K Potassium 39.102	20 Ca Calcium 40.08	21 Sc Scandium 44.956	22 Ti Titanium 47.90	23 V Vanadium 50.942	24 Cr Chromium 51.996	25 Mn Manganese 54.9380	26 Fe Iron 55.847	27 Co Cobalt 58.9332	28 Ni Nickel 58.71	29 Cu Copper 63.546	30 Zn Zinc 65.37	31 Ga Gallium 69.72	32 Ge Germanium 72.59	33 As Arsenic 74.9216	34 Se Selenium 78.96	35 Br Bromine 79.904	36 Kr Krypton 83.80
5	37 Rb Rubidium 85.47	38 Sr Strontium 87.62	39 Y Yttrium 88.905	40 Zr Zirconium 91.22	41 Nb Niobium 92.906	42 Mo Molybdenum 95.94	43 Tc Technetium (99)	44 Ru Ruthenium 101.07	45 Rh Rhodium 102.905	46 Pd Palladium 106.4	47 Ag Silver 107.868	48 Cd Cadmium 112.40	49 In Indium 114.82	50 Sn Tin 118.69	51 Sb Antimony 121.75	52 Te Tellurium 127.60	53 I Iodine 126.9044	54 Xe Xenon 131.30
6	55 Cs Cesium 132.905	56 Ba Barium 137.34	57 La Lanthanum 138.91	72 Hf Hafnium 179.49	73 Ta Tantalum 180.948	74 W Tungsten 183.85	75 Re Rhenium 186.2	76 Os Osmium 190.2	77 Ir Iridium 192.2	78 Pt Platinum 195.09	79 Au Gold 196.967	80 Hg Mercury 200.59	81 Tl Thallium 204.37	82 Pb Lead 207.19	83 Bi Bismuth 208.980	84 Po Polonium (210)	85 At Astatine (210)	86 Rn Radon (222)
7	87 Fr Francium (223)	88 Ra Radium (226)	89 Ac Actinium (227)	104 Rf Rutherfordium (261)	105 Db Dubnium (262)	106 Sg Seaborgium (266)	107 Bh Bohrium (264)	108 Hs Hassium (269)	109 Mt Meitnerium (268)	110 Ds Darmstadtium (269)	111 Rg Roentgenium (272)	112 Uub Ununbium (277)	113 Uut §	114 Uuq Ununquadium (285)	115 Uup §	116 Uuh Ununhexium (289)	117 Uus §	118 Uuo Ununoctium (293)

Lanthanide Series	58 Ce Cerium 140.12	59 Pr Praseodymium 140.907	60 Nd Neodymium 144.24	61 Pm Promethium (145)	62 Sm Samarium 150.35	63 Eu Europium 151.96	64 Gd Gadolinium 157.25	65 Tb Terbium 158.924	66 Dy Dysprosium 162.50	67 Ho Holmium 164.930	68 Er Erbium 167.26	69 Tm Thulium 168.934	70 Yb Ytterbium 173.04	71 Lu Lutetium 174.97
Actinide Series	90 Th Thorium 232.038	91 Pa Protactinium (231)	92 U Uranium 238.03	93 Np Neptunium (237)	94 Pu Plutonium (242)	95 Am Americium (243)	96 Cm Curium (247)	97 Bk Berkelium (247)	98 Cf Californium (251)	99 Es Einsteinium (254)	100 Fm Fermium (257)	101 Md Mendelevium (258)	102 No Nobelium (259)	103 Lr Lawrencium (260)

§ Note: Elements 113, 115, and 117 are not known at this time, but are included in the table to show their expected positions.

Starting on the far left of the periodic table is group IA (notice the group label atop each column), also known as the *alkali metals.* These most metallic of elements are very *reactive* (meaning they tend to combine with other elements) and are never found naturally in a pure state, but in a bonded state. Group IIA is called the *alkaline earth metals.* Just like the alkali metals, the alkaline earth metals are highly reactive.

The large central block of the periodic table is occupied by the *transition metals,* which are mostly listed as group B elements. Transition metals have properties that vary from extremely metallic, at the left side, to far less metallic, on the right side. The right-most boundary of the transition metals is shaped like a staircase, shown in bold in Figure 4-1.

✔ Elements to the left of the staircase are *metals* (except hydrogen in group IA).

✔ Elements to the right of the staircase are *nonmetals.*

✔ Elements bordering the staircase (boron, silicon, germanium, arsenic, antimony, tellurium, polonium, and astatine) are called *metalloids* because they have properties between those of metals and nonmetals. Chemists debate the membership of certain elements (especially polonium and astatine) within the metalloids, but the list given here reflects an inclusive view. You can find these elements in groups IIIA, IVA, VA, VIA, and VIIA.

Metals tend to be solid and shiny, conduct electricity and heat, give up electrons, and be *malleable* (easily shaped) and *ductile* (easily drawn out into wire). Nonmetals have properties opposite those of metals. The most extreme nonmetals are the *noble gases,* in group VIIIA on the far right of the table. The noble gases are *inert,* or extremely unreactive. One column to the left, in group VIIA, is another key family of nonmetals, the *halogens.* In nature, the reactive halogens tend to bond with metals to form salts, like sodium chloride (NaCl).

Be aware that some periodic tables use a different system for labeling the groups, in which each column is simply numbered from 1 to 18, left to right.

So, elements within a group tend to have similar properties. The similarities result from the fact that elements in a group have similar arrangements of electrons at their outermost border or *shell* (we describe electron shells in greater detail in the next section).

As you move across a period, however, the properties of the elements tend to change gradually. This slow change results mostly because the elements have different electron arrangements at their outermost shell. By the time you reach the right edge of the periodic table (group VIIIA), the elements' outermost shells are completely full. As we describe later, full outer shells mean that elements have no drive to react. The group VIIIA elements are called *noble gases* because they seem to consider themselves above the other elements; it would be beneath them to react.

Periods 6 and 7 have an added wrinkle. Elements with atomic numbers 58–71 and 90–103 are lifted out of the regular order and placed below the rest of the table. These two series are the *lanthanides* and the *actinides,* respectively. These elements are separated from the table for two main reasons. First, doing so prevents the table from being inconveniently wide. Second, the lanthanides all have pretty similar properties, as do the actinides.

Q. The elements within a group have widely varying numbers of protons, neutrons, and electrons. Why, then, do elements in a group tend to have similar chemical properties?

A. Chemical properties come mostly from the arrangement of electrons in the outermost shell of an atom. Although the elements at the top and bottom of a given group (like fluorine and astatine, for example) have very different numbers of protons, neutrons, and electrons, the arrangements of electrons in their outermost shells are very similar.

1. Are the following elements metals or non-metals?

   a. selenium

   b. fluorine

   c. strontium

   d. chromium

   e. bismuth

Solve It

2. Why are the noble gases referred to as "noble"?

Solve It

Predicting Properties from Periodic and Group Trends

The whole point of the periodic table, aside from providing interior decoration for chemistry classrooms, is to help predict and explain the properties of the elements. These properties change as a function of the numbers of protons and electrons in the element.

Increasing numbers of protons increase the positive charge of the nucleus, which contributes to *electron affinity,* the attraction an atom has for an added electron. Within a period, the more protons an element has, the stronger its electron affinity tends to be. This trend isn't perfectly smooth because other, more subtle factors are at work, but it's a good general description of what happens as you move across a period. Group VIIIA elements are an important exception. These elements have full outer shells and have no use for an added electron.

Increasing the number of electrons changes the reactivity of the element in predictable ways, based on how those electrons fill successive energy levels. Electrons in the highest energy level occupy the outermost shell of the atom and are called *valence electrons.* Valence electrons determine whether an element is reactive or unreactive and are the electrons involved in bonding (which we talk about in Chapter 5). Because chemistry is really about the making and breaking of bonds, valence electrons are the most important particles for chemistry.

Atoms are most stable when their valence shells are completely filled with electrons. Chemistry happens as a result of atoms attempting to fill the valence shells. Alkali metals in group IA of the periodic table are so reactive because they need only give up one electron to have a completely filled valence shell. Halogens in group VIIA are so reactive because they need only acquire one electron to have a completely filled valence shell. The elements within a group tend to have the same number of valence electrons, and for that reason tend to have similar chemical properties. Elements in groups IA and IIA tend to react strongly with elements in group VIIA. Group B elements tend to react less strongly.

Elements in the A groups have the same number of valence electrons as the Roman numeral of their group. For example, magnesium in group IIA has two valence electrons.

In addition to reactivity, another property that varies across the table is *atomic radius,* or the geometric size (not the mass) of the atoms. As you move down or to the right on the periodic table, elements have both more protons and more electrons. However, only as you move down the table do the added electrons occupy higher energy levels. So, atomic radius tends to decrease as you move to the right because the increasing positive charge of the nuclei pulls inwardly on the electrons of that energy level. As you move down the table, atomic radius tends to increase because, even though you are adding positively charged protons to the nucleus, you are now adding electrons to higher and higher energy levels, which correspond to larger and larger electron shells. So you can make good estimates about the relative sizes of the radii of different elements by comparing their placement in the periodic table.

Q. Chromium has more electrons than scandium. Why, then, does scandium have a larger atomic radius?

A. Even though chromium has more electrons than scandium, those extra electrons occupy the same energy level because the two elements are in the same row on the periodic table. Furthermore, chromium has more protons than scandium (as you can tell by chromium's position to the right of scandium), creating a more positively charged nucleus that pulls the electrons of a given energy level inward, thereby decreasing the atomic radius.

3. Which has a larger atomic radius, silicon or barium?

Solve It

4. Which has a stronger electron affinity, silicon or barium?

Solve It

5. How many valence electrons do the following elements have?

  a. I

  b. O

  c. Ca

  d. H

  e. Ge

Solve It

6. Why are valence electrons important?

Solve It

Seeking Stability with Valence Electrons by Forming Ions

Elements are so insistent about having filled valence shells that they'll gain or lose valence electrons to do so. Atoms that gain or lose electrons in this way are called *ions*. You can predict what kind of ion many elements will form simply by looking at their position on the periodic table. With the exception of row 1 (hydrogen and helium), all elements are most stable (think "happiest") with a full shell of eight valence electrons, known as an *octet*. Atoms tend to take the shortest path to a complete octet, whether that means ditching a few electrons to achieve a full octet at a lower energy level, or grabbing extra electrons to complete the octet at their current energy level. In general, metals on the left side of the periodic table (and in the middle) tend to lose electrons, and nonmetals on the right tend to gain electrons.

More specifically, you can predict just how many electrons an atom will gain or lose to become an ion. Group IA (alkali metal) elements lose one electron. Group IIA (alkaline earth metal) elements lose two electrons. Things get unpredictable in the transition metal region, but get predictable once more with the nonmetals. Group VIIA (halogen) elements gain one electron. Group VIA elements tend to gain two electrons, and group VA elements tend to gain three electrons. In short, elements tend to lose or gain as many electrons as necessary to have valence shells resembling the elements in group VIIIA, the noble gases.

When atoms become ions, they lose the one-to-one balance between their protons and electrons, and therefore acquire an overall charge.

> ✔ Atoms that lose electrons (like metals) acquire positive charge, becoming *cations*, such as Na^+ or Mg^{2+}.

> ✔ Atoms that gain electrons (like nonmetals) acquire negative charge, becoming *anions*, such as Cl^- or O^{2-}.

The superscripted numbers and signs in the atoms' symbols indicate the ion's overall charge. Cations have superscripts with "+" signs, and anions have superscripts with "−" signs. When the element sodium, Na, loses an electron, it loses one negative charge and is left with one overall positive charge. So, Na becomes Na^+.

Q. Fluorine and sodium are only two atomic numbers apart on the periodic table. Why then does fluorine form an anion, F^-, whereas sodium forms a cation, Na^+?

A. Fluorine (F) is in group VIIA, just one group to the left of the noble gases, and therefore needs to gain only one electron to complete a valence octet. Sodium (Na) lies just one group to the right of the noble gases, having wrapped around into group IA of the next energy level. Therefore, sodium needs to lose only one electron to achieve a full valence octet.

---

7. How many electrons will be gained or lost by the following elements when forming an ion?

a. Lithium

b. Selenium

Solve It

8. What type of ion is nitrogen most likely to form?

Solve It

9. What type of ion is beryllium most likely to form?

Solve It

Putting Electrons in Their Places: Electron Configurations

An awful lot of detail goes into determining just how many electrons an atom has (see the previous sections). The next step is to figure out where those electrons live. Several different schemes exist for annotating all of this important information, but the *electron configuration* is a type of shorthand that captures much of the pertinent information.

Each numbered row of the periodic table corresponds to a different *principal energy level,* with higher numbers indicating higher energy. Within each energy level, electrons can occupy different *orbitals*. Different types of orbitals have slightly different energy levels. Each orbital can hold up to two electrons, but electrons won't double up within an orbital unless no other open orbitals exist at the same energy level. Electrons fill up orbitals from the lowest energies to the highest, as the rest of us would probably do if we were electrons.

There are four types of orbitals: *s, p, d,* and *f.*

- ✔ Row 1 consists of a single 1*s* orbital. A single electron in this orbital corresponds to the electron configuration of hydrogen, written as $1s^1$. The superscript written after the symbol for the orbital indicates how many electrons occupy that orbital. Filling the orbital with two electrons, $1s^2$, corresponds to the electron configuration of helium. Each higher principal energy level contains its own s orbital (2*s*, 3*s*, and so on), and these orbitals are the first to fill within those levels.

- ✔ In addition to *s* orbitals, principal energy levels 2 and higher contain *p* orbitals. There are three *p* orbitals at each level, accommodating a maximum of six electrons. Because the three *p* orbitals (also known as p_x, p_y, and p_z) have equal energy, they are each filled with a single electron before any receives a second electron. The elements in rows 2 and 3 on the periodic table contain only *s* and *p* orbitals. The *p* orbitals of each energy level are filled only after the *s* orbital is filled.

- ✔ Rows 4 and higher on the periodic table include *d* orbitals, of which there are five at each principal energy level, accommodating a maximum of ten electrons. Rows 5 and higher include *f* orbitals, numbering seven at each level, accommodating a maximum of 14 electrons. *d* orbital electrons are a major feature of the transition metals. *f* orbital electrons are a hallmark of the lanthanides and the actinides (see the "Reading Periods and Groups in the Periodic Table" section, earlier in this chapter, for more on these rows).

After you get to row 4, the exact order in which you fill the energy levels can get a bit confusing. To keep things straight, it's useful to refer to the Aufbau filling diagram, shown in Figure 4-2. To use the diagram, start at the bottom and work your way up, from the lowest arrows to the highest. For example, always start by filling 1*s*, then fill 2*s*, then 2*p*, then 3*s*, then 4*s*, then 3*d*, and so on.

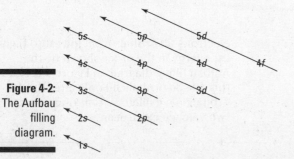

Figure 4-2:
The Aufbau
filling
diagram.

Sadly, there are a few exceptions to the tidy picture presented by the Aufbau filling diagram. Copper, chromium, and palladium are notable examples (see Chapter 25 for more information on these). Without going into teeth-grinding detail, these exceptional electron configurations arise from situations where electrons get transferred from their proper, Aufbau-filled orbitals to create half-filled or entirely filled sets of *d* orbitals; these half- and entirely filled states are more stable than the states produced by pure Aufbau-based filling.

To come up with a written electron configuration, you first determine how many electrons the atom in question actually has. Then you assign those electrons to orbitals, one electron at a time, from the lowest energy orbitals to the highest. In a given type of orbital (like a 2*p* or 3*d* orbital, for example), you only place two electrons within the same orbital when there is no other choice. For example, suppose you want to find the electron configuration of carbon. Carbon has six electrons, the same as the number of its protons as described by its atomic number (see Chapter 3 for more about atomic numbers), and it's in row 2 of the periodic table. First, the *s* orbital of level 1 is filled. Then, the *s* orbital of level 2 is filled. These orbitals each accept two electrons, leaving two more with which to fill the *p* orbitals of level 2. Each of the remaining electrons would occupy a separate *p* orbital. You wind up with $1s^2 2s^2 2p^2$. (Only at oxygen, $1s^2 2s^2 2p^4$, would electrons begin to double up in the 2*p* orbitals, indicated by the superscript 4 on the *p*.)

Electron configurations can get a bit long to write. For this reason, you may sometimes see them written in a condensed form, like $[Ne]3s^2 3p^3$. This condensed configuration is the one for phosphorus. The expanded electron configuration for phosphorus is $1s^2 2s^2 2p^6 3s^2 3p^3$. To abbreviate the configuration, simply go backward in atomic numbers until you hit the nearest noble gas (which, in the case of phosphorus, is neon). That symbol for that noble gas, laced within brackets, becomes the new starting point for the configuration. Next, simply include the standard configuration for those electrons beyond the noble gas (which, in the case of phosphorus, includes two electrons in 3*s* and three electrons in 3*p*). The condensed form simply means that the atom's electron configuration is just like that of neon, with additional electrons filled into orbitals 3*s* and 3*p* as annotated.

Ions have different electron configurations than their parent atoms because ions are created by gaining or losing electrons. As you find out in the previous section, atoms tend to gain or lose electrons so they can achieve full valence octets, like those of the noble gases. Guess what? The resulting ions have precisely the same electron configurations as those noble gases. So, by forming the Br⁻ anion, bromine achieves the same electron configuration as the noble gas, krypton.

Q. What is the electron configuration of titanium?

A. $1s^22s^22p^63s^23p^64s^23d^2$. Titanium (Ti) is atomic number 22, and therefore has 22 electrons to match its 22 protons. These electrons fill orbitals from lowest to highest energy in the order shown by the Aufbau filling diagram in Figure 4-2. Note that the 3d orbitals fill only after the 4s orbitals have filled, so that titanium has two valence electrons.

---

10. What is the electron configuration for chlorine?

11. What is the electron configuration for technetium?

---

12. What is the condensed electron configuration of bromine?

13. What are the electron configurations of chlorine anion, argon, and potassium cation?

Measuring the Amount of Energy (Or Light) an Excited Electron Emits

Electrons can jump between energy levels by absorbing or releasing energy. When an electron absorbs an amount of energy exactly equivalent to the difference in energy between levels (a *quantum* of energy), the electron jumps to the higher energy level. This jump in energy creates an *excited state*. Excited states don't last forever; the lower-energy *ground state* is more stable. Excited electrons tend to release amounts of energy equivalent to the difference between energy levels, and thereby return to ground state. The discrete particles of light that correspond to quanta of energy are called *photons*.

Light has properties of both a particle and a wave. A lovely result arises from light's properties: You can measure differences in electron energies simply by measuring the wavelengths of light emitted from excited atoms. In this way, you can identify different elements within a sample. The first basic relationship behind this technique is

> Speed of light (c) = Wavelength (λ) × Frequency (ν)

where c = 3.00×10^8 m s^{-1}. Frequency, ν, is often expressed in reciprocal seconds (1/s or s^{-1}), also known as hertz (1 Hz = 1 s^{-1}).

But what about energy? What's the relationship between frequency and energy? It must be ridiculously complicated, right? Wrong. The second basic relationship relating light to energy is

> Energy (E) = Planck's constant (h) × Frequency (ν)

where Planck's constant, h = 6.626×10^{-34} J · s.

Here, J stands for *joules,* the International System (SI) unit of energy. Frequency is expressed in hertz (Hz), where 1 hertz is 1 inverse second (s^{-1}). So, multiplying a frequency by Planck's constant yields *joules,* the units of energy.

Q. A hydrogen lamp emits blue light at a wavelength of 487 nanometers (nm). What is the frequency in Hertz?

A. 6.16×10^{14} s^{-1}. Wavelength and the speed of light are known quantities here; you must solve for frequency, ν = c / λ. Don't forget to convert nanometers to meters (1 nm = 10^{-9}m), because the speed of light is given in meters per second:

$$\frac{3.00 \times 10^8 \text{m s}^{-1}}{487 \times 10^{-9} \text{m}} = 6.16 \times 10^{14} \text{s}^{-1}$$

Q. What is the energy of emitted light with a frequency of 6.88×10^{14} Hz?

A. 4.56×10^{-19} J. E = hν = $(6.626 \times 10^{-34}$ J · s) $\times (6.88 \times 10^{14}$ s$^{-1}) = 4.56 \times 10^{-19}$ J. You must recall here that Hz = s^{-1}.

$$E = \left(6.626 \times 10^{-34} \text{J} \cdot \text{s}\right)\left(6.88 \times 10^{14} \text{s}^{-1}\right) =$$

$$4.56 \times 10^{-19} \frac{\text{J} \cdot \text{s}}{\text{s}} = 4.56 \times 10^{-19} \text{J}$$

14. What is the frequency of a beam of light with wavelength 2.57×10^2 m?

 Solve It

15. One useful, high-energy emission from excited electrons is the X-ray. What is the wavelength (in meters) of an X-ray with $\nu = 8.72 \times 10^{17}$ Hz?

Solve It

16. What is the energy level corresponding to an emission at 3.91×10^8 Hz?

Solve It

Answers to Questions on the Periodic Table

Here are the answers to the practice problems in this chapter. Pat yourself on the back, whether in congratulation or consolation.

1 Metals are found to the left of the staircase, with the exception of hydrogen, and nonmetals are found to the right of the staircase. Several elements adjoining the staircase are classified as metalloids.

    a. selenium (Se) = **nonmetal**

    b. fluorine (F) = **nonmetal**

    c. strontium (Sr) = **metal**

    d. chromium (Cr) = **metal**

    e. bismuth (Bi) = **metal**

2 The noble gases are described as "noble" because they seem to consider it beneath themselves to react with other elements. Because these elements have completely filled valence shells, they have no energetic reason to react. They're already as stable as they can be.

3 **Barium's atomic radius is larger.** Atomic size tends to increase from top to bottom and from right to left. Barium (Ba) is farther down and to the left on the periodic table than silicon (Si).

4 **Silicon has stronger electron affinity.** Electron affinity tends to increase from left to right.

5 When you're looking at elements in A groups, the number of valence electrons matches the element's group number.

    a. Iodine (I) has **seven valence electrons** because it's in group VIIA.

    b. Oxygen (O) has **six valence electrons** because it's in group VIA.

    c. Calcium (Ca) has **two valence electrons** because it's in group IIA.

    d. Hydrogen (H) has **one valence electron** because it's in group IA.

    e. Germanium (Ge) has **four valence electrons** because it's in group IVA.

6 Valence electrons are important because they occupy the highest energy level in the outermost shell of the atom. As a result, valence electrons are the electrons that see the most action, as in forming bonds, or being gained or lost to form ions. The number of valence electrons in an atom largely determines the chemical reactivity of that atom.

7 **a. Lithium (Li) loses one electron,** forming lithium cation Li^+.

    b. Selenium (Se) gains two electrons to form selenium anion Se^{2-}.

Elements seek stable valence shells like those of the noble gases by gaining or losing electrons. Whether an element gains or loses electrons has to do with where that elements sits in the periodic table. For elements on the left side, losing just a few electrons is easier than gaining many more; for these elements, the Roman numeral of the A group to which they belong tells you how many electrons they lose — and therefore tells you their positive charge. For elements on the right side of the table, gaining a few electrons is easier than losing many more; these elements gain the number of electrons necessary to create valence shells like those in group VIIIA (the noble gases). An element in group VIA therefore gains two electrons.

8 N^{3-}, also known as nitride or nitrogen anion. By gaining three electrons, nitrogen, which is in group VA, assumes a full octet, like neon.

9 Be^{2+}, also known as beryllium cation. By losing two electrons, beryllium, which is in group IIA, assumes a full octet valence shell, with two electrons like helium.

10 $1s^2 2s^2 2p^6 3s^2 3p^5$. Chlorine (Cl) has 17 electrons.

11 $1s^2 2s^2 2p^6 3s^2 3p^6 4s^2 3d^{10} 4p^6 5s^2 4d^5$. Technetium (Tc) has 43 electrons, which fill according to the Aufbau diagram.

12 $[Ar]4s^2 3d^{10} 4p^5$. Condense the configuration for bromine (Br) by summarizing all previous, entirely filled rows with a single noble gas in brackets. The expanded electron configuration of bromine is $1s^2 2s^2 2p^6 3s^2 3p^6 4s^2 3d^{10} 4p^5$. Bromine is in row 4, so consolidate the electron configurations of rows 1 through 3 as [Ar], because argon is the last element in row 3. Then add to [Ar] the configuration of the remaining electrons, which fill $4s^2 3d^{10} 4p^5$.

13 **All three have the configuration $1s^2 2s^2 2p^6 3s^2 3p^6$**, the configuration of the noble gas argon (Ar). Both chlorine (Cl) and potassium (K) seek stable, full octet configurations by forming ions.

14 $11.7 \times 10^5 \text{ s}^{-1}$.

$$\frac{3.00 \times 10^8 \text{m s}^{-1}}{2.57 \times 10^2 \text{m}} = 11.7 \times 10^5 \text{s}^{-1}$$

15 3.44×10^{-10} m.

$$\frac{3.00 \times 10^8 \text{m s}^{-1}}{8.72 \times 10^{17} \text{s}^{-1}} = 3.44 \times 10^{-10} \text{m}$$

16 2.59×10^{-25} J.

$$\left(6.626 \times 10^{-34} \text{J} \cdot \text{s}\right)\left(3.91 \times 10^8 \text{s}^{-1}\right) = 2.59 \times 10^{-25} \text{J}$$

Part II
Making and Remaking Compounds

The 5th Wave   By Rich Tennant

"Okay—now that the paramedic is here with the defibrillator and smelling salts, prepare to learn about covalent bonds."

In this part . . .

Chemistry begins when elements start making friends. A compound is a group of elemental friends. Some friends are closer than others. Potential new friends test old alliances, sometimes breaking them to form new ones. In other words, reactions occur. This part investigates compounds and bonds, the forces that hold compounds together. Because the friendly population of elements and compounds is so large, we introduce the concept of the mole, a really big number that makes counting countless compounds infinitely easier. We also introduce you to reactions, the dramatic dialogue of elements and compounds that underlies all chemical change.

Chapter 5

Building Bonds

· ·

In This Chapter

▶ Giving and receiving electrons in ionic bonding

▶ Sharing electrons in covalent bonding

▶ Understanding molecular orbitals

▶ Shaping up molecules with VSEPR theory and hybridization

▶ Tugging at the idea of polarity

· ·

Many atoms are prone to public displays of affection, pressing themselves against other atoms in an intimate electronic embrace called *bonding*. Atoms bond with one another by playing various games with their valence electrons. In this chapter, we describe the basic rules of those games.

Because valence electrons are so important to bonding, problems involving bonding sometimes make use of *electron dot structures,* symbols that represent valence electrons as dots surrounding an atom's chemical symbol. You should be able to draw and interpret electron dot structures for atoms as shown in Figure 5-1. This figure shows the electron dot structures for elements in the periodic table's first two rows; notice that the valence shells progressively fill moving from left to right. To determine the electron dot structure of any element, count the number of electrons in that element's valence shell. Then draw that number of dots around the chemical symbol for the element. Chapter 4 describes some of the factors that determine whether atoms gain or lose electrons to form ions. You should make sure to understand those patterns before attacking this chapter.

Figure 5-1:
Electron dot
structures
for elements
in the first
two rows of
the periodic
table.

IA	IIA	IIIA	IVA	VA	VIA	VIIA	VIIIA
H ·							He :
Li ·	·Be·	·Ḃ·	·C̈·	:N̈:	:Ö:	:F̈:	:N̈e:

Pairing Charges with Ionic Bonds

Atoms of some elements, like metals, can easily lose valence electrons to form *cations* (atoms with positive charge) that have stable electron configurations. Atoms of other elements, like the halogens, can easily gain valence electrons to form *anions* (atoms with negative charge) with stable electron configurations. Cations and anions experience *electrostatic attraction* to one another because opposite charges attract. So, a cation will snuggle up to an anion, given the chance. This event is called *ionic bonding*, and it happens because the energy of the ionically bonded ions is lower than the energy of the ions when they are separated.

You can think of an ionic bond as resulting from the transfer of an electron from one atom to another, as shown in Figure 5-2 for sodium and chlorine. Metals (like sodium) tend to give up their electrons to nonmetals (like chlorine) because nonmetals are much more *electronegative* (they more strongly attract electrons within a bond to themselves). The greater the difference in electronegativity between the two ions, the more *ionic* (or completely uneven in sharing of electrons) is the bond that forms between them.

Figure 5-2:
The transfer of an electron from sodium to chlorine to form an ionic bond between the Na^+ cation and the Cl^- anion.

$$Na \cdot \longrightarrow \cdot \ddot{C}\ddot{l}\colon \longrightarrow Na^+ \quad \colon\!\ddot{C}\ddot{l}\colon^-$$

Although ions are often individual, charged atoms, there are also many examples of *polyatomic ions* (charged particles made up of more than one atom). Examples of common polyatomic ions are ammonium, NH_4^+, and sulfate, SO_4^{2-}. We cover polyatomic ions in detail in Chapter 6.

When cations and anions associate in ionic bonds, they form *ionic compounds*. At room temperature, most ionic compounds are solid because of the strong electrostatic forces that hold together the ions within them. The ions in ionic solids tend to pack together in a *lattice*, a highly organized, regular arrangement that allows for the maximum electrostatic interaction between anions and cations. The geometric details of the packing can differ among different ionic compounds, but a simple lattice structure is shown in Figure 5-3. Flip to Chapter 6 for full details on ionic compounds.

The strong electrostatic forces that hold together ionic lattices result in the high melting and boiling points that are common among ionic compounds (see Chapter 10 for general information on melting and boiling points). Although it may take a great deal of thermal energy to disrupt ionic bonds, ionic compounds are usually easily dissolved in water or in other *polar solvents* (fluids made up of molecules that have unevenly distributed charge). When the solvent molecules are polar, they can engage in favorable interactions with the ions that help to compensate for disrupting the ionic bonds. For example, polar water molecules can interact well with both sodium cations (Na^+) and chlorine anions (Cl^-). Water molecules are polar because they have distinct and separate bits of positive and negative charge. Water molecules can orient their positive bits toward Cl^- and their negative bits toward Na^+. Positive charges attract negative charges and vice versa, so these kinds of interactions are favorable — they require less energy. So, water dissolves solid NaCl quite well because the water-ion interactions can compete with the (Na^+)-(Cl^-) interactions.

Figure 5-3:
The lattice structure of an ionic solid, sodium chloride.

Na^+
Cl^-

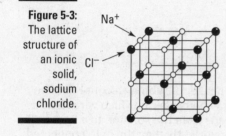

When ionic compounds are melted or dissolved, so the individual ions can move about, the resulting liquid is a very good conductor of electricity. Ionic solids, however, are often poor conductors of electricity.

Salts are a common variety of ionic compound. A salt is formed from the reaction between a base and an acid. For example, hydrochloric acid reacts with sodium hydroxide to form the salt sodium chloride and water:

$$HCl(aq) + NaOH(aq) \rightarrow NaCl(aq) + H_2O(l)$$

Note that *aq* indicates that the substance is dissolved in water, in an *aqueous* solution.

Q. Why do metals tend to form ionic compounds with nonmetals?

A. Metals are much less electronegative than nonmetals, meaning that they give up valence electrons much more easily. Nonmetals (especially group VIIA and VIA nonmetals) very easily gain new valence electrons. So, metals and nonmetals tend to form bonds in which the metal atoms entirely surrender valence electrons to the nonmetals. Bonds with extremely unequal electron-sharing are called ionic bonds.

1. What is the electron dot structure of potassium fluoride?

Solve It

2. The ionic compound lithium sulfide forms between the elements lithium and sulfur. In which direction are electrons transferred to form ionic bonds, and how many electrons are transferred?

Solve It

3. Magnesium chloride is dissolved into a beaker of water and a beaker of rubbing alcohol until no more compound will dissolve. Electrical circuits are set up for each beaker in which wires lead from a battery into the solution, and a separate set of wires leads from the solution to a light bulb. The bulb connected to the aqueous solution circuit glows more brightly than the bulb connected to the alcohol solution circuit. Why?

Solve It

Sharing Charge with Covalent Bonds

Sometimes the way for atoms to reach their most stable, lowest-energy states is to share valence electrons. When atoms share valence electrons, we say that they are engaged in *covalent bonding*. The very word *covalent* means "together in valence." Compared to ionic bonding, covalent bonding tends to occur between atoms of similar electronegativity, most especially between nonmetals.

Just as ionic bonds tend to form in such a way that both atoms end up with completely filled valence shells, the atoms involved in covalent bonds tend to share electrons in such a way that each ends up with a completely filled valence shell. The shared electrons are attracted to the nuclei of both atoms, forming the bond. The simplest and best studied covalent bond is the one formed between two hydrogen atoms, shown in Figure 5-4. Separately, each atom has only one electron with which to fill its $1s$ orbital. By forming a covalent bond, each atom lays claim to two electrons within the molecule of dihydrogen. The figure shows various ways in which a covalent bond can be represented, explicitly depicting the valence shells (a), by using electron dot structures (b), or by signifying a shared pair of electrons with a single line (c). The latter two ways to show bonding are referred to as *Lewis structures*.

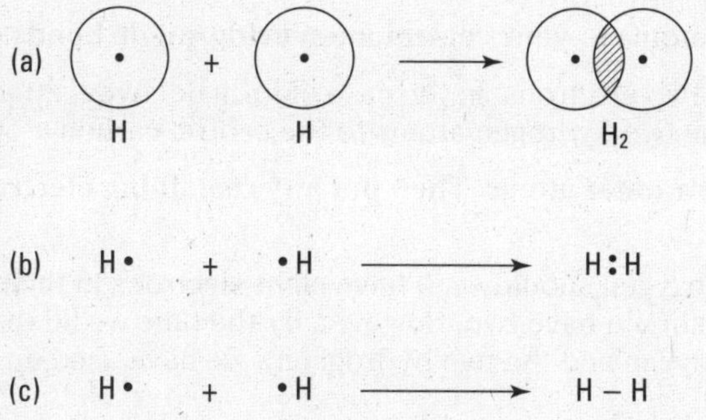

Figure 5-4:
Three representations of the formation of a covalent bond in dihydrogen.

Atoms can share more than a single pair of electrons. When atoms share two pairs of electrons, they are said to form a *double bond,* and when they share three pairs of electrons they are said to form a *triple bond.* Examples of double and triple bonds are shown with electron dot and line structures in Figure 5-5.

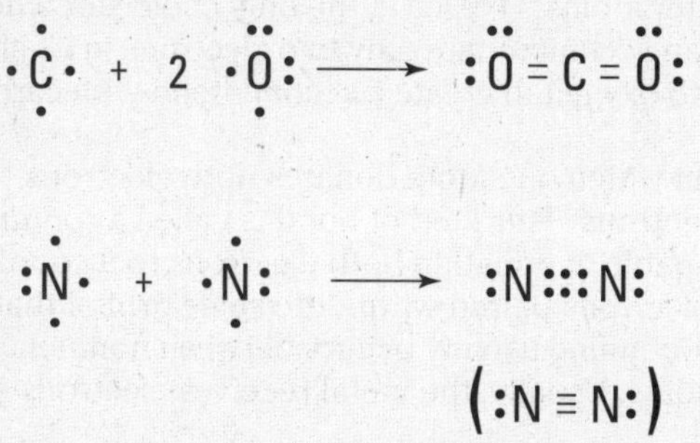

Figure 5-5:
The formation of double bonds in carbon dioxide and triple bonds in dinitrogen.

A few guidelines can help you figure out the correct Lewis structure for a molecule if you know the molecule's formula. As an example, we work out the Lewis structure of formaldehyde, CH_2O (Figure 5-6 can help you follow along):

1. **Add up all the valence electrons for all the atoms in the molecule.**

   These are the electrons you can use to build the structure. Account for any extra or missing electrons in the case of ions. For example, if you know your molecule has +2 charge, remember to subtract two from the total number of valence electrons. In the case of formaldehyde, C has four valence electrons, each H has one valence electron, and O has six valence electrons. The total number of valence electrons is 12.

2. **Pick a "central" atom to serve as the anchor of your Lewis structure.**

   The central atom is usually one that can form the most bonds, which is often the atom with the most empty valence orbital slots to fill. In larger molecules, some trial-and-error may be involved in this step, but in smaller molecules, some choices are obviously better than others. For example, carbon is a better choice than hydrogen to be the central atom because carbon tends to form four bonds, whereas hydrogen tends to form only one bond. In the case of formaldehyde, carbon is the obvious first choice because it can form four bonds, while oxygen can form only two, and each hydrogen can form only one.

3. **Connect the other, "outer" atoms to your central atom using single bonds only.**

Each single bond counts for two electrons. In the case of formaldehyde, attach the single oxygen and each of the two hydrogen atoms to the central carbon atom.

4. **Fill the valence shells of your outer atoms. Then put any remaining electrons on the central atom.**

In our example, carbon and oxygen should each have eight electrons in their valence shells; each hydrogen atom should have two. However, by the time we fill the valence shells of our outer atoms (oxygen and the two hydrogens), we have used up our allotment of 12 electrons.

5. **Check whether the central atom now has a full valence shell.**

If the central atom has a full valence shell, then your Lewis structure is drawn properly — it's formally correct even though it may not correspond to a real structure. If the central atom still has an incompletely filled valence shell, then use electron dots (nonbonding electrons) from outer atoms to create double and/or triple bonds to the central atom until the central atom's valence shell is filled. Remember, each added bond requires two electrons. In the case of our formaldehyde molecule, we must create a double bond between carbon and one of the outer atoms. Oxygen is the only choice for a double-bond partner, because each hydrogen can accommodate only two electrons in its shell. So, we use two of the electrons assigned to oxygen to create a second bond with carbon.

Sometimes a covalent bond is formed in which one atom donates both electrons to the bond, with the other atom contributing no electrons. This kind of bond is called a *coordinate covalent bond.* Atoms with lone pairs are capable of donating both electrons to a coordinate covalent bond. A *lone pair* consists of two electrons paired within the same orbital that aren't used in bonding. Even though covalent bonding usually occurs between nonmetals, metals can engage in coordinate covalent bonding. Usually, the metal receives electrons from an electron donor called a *ligand.*

1. $C(4 e^-) + H(1 e^-) + H(1 e^-) + O(6 e^-) = 12 e^-$

2. Carbon is central atom; it can form more bonds (4) than O, H.

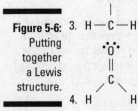

Figure 5-6: Putting together a Lewis structure.

Sometimes a given set of atoms can covalently bond with each other in multiple ways to form a compound. This situation leads to something called *resonance.* Each of the possible bonded structures is called a *resonance structure.* The actual structure of the compound is a *resonance hybrid,* a sort of average of all the resonance structures. For example, if two atoms are connected by a single bond in one resonance structure, and the same two atoms are connected by a double bond in a second resonance structure, then those atoms are connected by a bond in the resonance hybrid that is worth 1½ single bonds. A common example of resonance is found in ozone, O_3, shown in Figure 5-7.

Figure 5-7: Resonance structures of ozone, shown in two representations.

$$: \ddot{O} : \ddot{O} :: \ddot{O} : \qquad \longleftrightarrow \qquad : \ddot{O} :: \ddot{O} : \ddot{O} :$$
$$\;\;\;\;\;\; - \;\; + \qquad\qquad\qquad\qquad\qquad + \;\; -$$

or

$$: \ddot{O} - \ddot{O} = \ddot{O} : \qquad \longleftrightarrow \qquad : \ddot{O} = \ddot{O} - \ddot{O} :$$
$$\;\;\;\;\;\; - \;\; + \qquad\qquad\qquad\qquad\qquad + \;\; -$$

EXAMPLE

Q. Draw a Lewis structure for propene, C_3H_6.

A. First, add up the total valence electrons. Each carbon contributes 4 electrons, and each hydrogen contributes 1, for a total of 18 valence electrons. Next, pick a central atom. The best choice is a carbon atom because carbon can form four bonds, more than any hydrogen. Connect the remaining atoms to the central carbon with single bonds. To connect all the atoms into one molecule, the central carbon must be connected to each of the two other carbon atoms. These connections use up 16 of the 18 valence electrons, leaving 2 electrons that you can place onto one of the carbon atoms. One carbon atom in the structure still requires two additional electrons to fill its valence shell. The only way to fill this shell is to create a carbon-carbon double bond. Only one arrangement of hydrogen atoms to the three carbons allows you to fill all the carbon valence shells, as you can see in the following figure:

4. Bertholite is the common name for dichlorine, a toxic gas that has been used as a chemical weapon. Why is bertholite most certainly a covalently bonded compound? What is the most likely electron dot structure of this compound?

Solve It

5. When aluminum chloride salt is dissolved in water, aluminum (III) cations become surrounded by clusters of six water molecules to form a "hexahydrated" aluminum cation, $Al(H_2O)_6{}^{3+}$. Being a group IIIA metal, aluminum easily gives up its valence electrons. The oxygen atom in water possesses two lone pairs. What kind of bonding most likely occurs between the aluminum and the hydrating water molecules?

Solve It

6. Benzene, C_6H_6, is a common industrial solvent. The benzene molecule is based on a ring of covalently bonded carbon atoms. Draw two acceptable Lewis structures for benzene. Based on the structures, describe a likely resonance hybrid structure for benzene.

Solve It

Occupying and Overlapping Molecular Orbitals

Chapter 4 describes how electrons occupy distinct orbitals within atoms. When atoms covalently bond to form molecules, the shared electrons are no longer constrained to those atomic orbitals, but occupy *molecular orbitals,* larger regions that form from the overlap of atomic orbitals. Just as different atomic orbitals are associated with different levels of energy, so are molecular orbitals. A stable covalent bond forms between two atoms because the energy of the molecular orbital associated with the bond is lower than the combined energies associated with the atomic orbitals of the separated atoms.

Because electrons have wave-like properties, atomic orbitals can overlap in different ways depending on the relationship between the waves of the shared electrons.

 ✔ In one mode, the electron waves interact *favorably* (with low energy) and together occupy a *bonding orbital.*

 ✔ In another mode, the electron waves interact unfavorably within a higher-energy *antibonding orbital.*

The energy relationships between unbound atoms and different types of molecular orbitals are summarized within *molecular orbital diagrams,* such as the one shown for dihydrogen in Figure 5-8. In this figure, two hydrogen atoms each contribute a single electron from a 1s orbital to a sigma (σ) bonding orbital. The low-energy bonding orbital is favored over the higher-energy sigma antibonding (σ*) orbital. This illustrates a general principle: Given a choice between high- and low-energy states, molecules prefer the low-energy states. This preference for lower energy is what is meant by *favorable* (low energy) versus *unfavorable* (high energy).

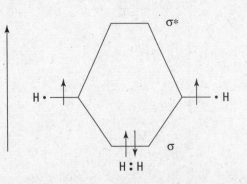

Figure 5-8:
A molecular orbital diagram for the formation of dihydrogen.

In addition to differences in the interaction of electron waves, covalent bonds can differ based on the shape of the molecular orbitals.

✔ When atomic orbitals overlap in such a way that the resulting molecular orbital is symmetrical with respect to the *bond axis* (the line connecting the two bonded atoms), we say that a σ bond *(sigma bond)* is formed.

✔ When atomic orbitals overlap in such a way that the resulting molecular orbital is symmetric with the bond axis in only one plane, we say that a π bond *(pi bond)* is formed.

Sigma bonds are stronger than pi bonds because the electrons within sigma bonds lie directly between the two atomic nuclei. The negatively charged electrons in sigma bonds therefore experience favorable (as in, low-energy) attraction to the positively charged nuclei. Electrons in pi bonds are farther away from the nuclei, so they experience weaker attraction.

Sigma bonds form when *s* or *p* orbitals overlap in a head-on manner. Single bonds are usually sigma bonds. Pi bonds form when adjacent *p* orbitals overlap above and below the bond axis. These situations are depicted in Figure 5-9.

Figure 5-9: Formation of a sigma bond (σ) from two *s* orbitals, and formation of a pi bond (π) from two adjacent *p* orbitals.

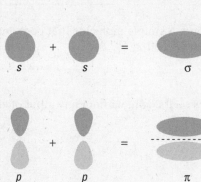

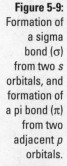

Q. Both sigma bonds and pi bonds have a kind of symmetry with respect to the two atoms in the bond. What is the difference in a sigma bond's symmetry versus a pi bond's symmetry?

A. In any bond between two atoms, you can imagine an imaginary line (the bond axis) that connects the center of one atom to the center of the other atom. Sigma bonds are perfectly symmetrical around this line; you can imagine the sigma bond as a kind of tube that wraps around the bond axis. If you were to rotate the bonded atoms around the imaginary line, the bond would look the same all the way around. Pi bonds are symmetric to the bond axis *in only one plane*. You can imagine the two atoms pressed onto a flat surface; the bond axis is the imaginary line on this surface. The pi bonds connect the two atoms above the line and below the line. If you were to rotate the bonded atoms around the imaginary line, the pi bonds would rise up off the surface and sink below the surface as you rotated them, like the planks on a paddlewheel rise above the surface of the water and then sink below the surface of the water.

7. Draw a molecular orbital diagram for the hypothetical molecule, dihelium.

Solve It

8. Based on the molecular orbital diagram of dihelium, explain why dihelium is far less likely to exist than dihydrogen.

Solve It

9. Double bonds involve one sigma bond and one pi bond. A simple molecule that contains a double bond is ethene, $H_2C=CH_2$. Ethene reacts with water to form ethanol:

$H_2C=CH_2 + H_2O \rightarrow H_3C-COH$

This reaction is favorable, meaning that it progresses on its own, without any input of energy. Why might this be the case?

Solve It

Tugging on Electrons within Bonds: Polarity

The earlier sections on ionic bonding and covalent bonding refer to the concept of electronegativity, or the tendency of an atom to draw electrons toward itself. Ionic bonds form between atoms with large differences in electronegativity, whereas covalent bonds form between atoms with smaller differences in electronegativity. In truth, there is no natural distinction between the two types of bonds; they lie on opposite sides of a spectrum of *polarity,* or unevenness in the distribution of electrons within a bond.

The greater the difference in electronegativity between two atoms, the more polar is the bond that forms between them. Imagine the electrons in the bond as being spread out into a cloud within the molecular orbital. In polar bonds, the cloud is denser in the vicinity of the more electronegative atom. In nonpolar bonds, like those formed between atoms of the same element, the cloud is evenly distributed between both atoms. Polar bonds have more ionic character, whereas nonpolar bonds have more covalent character. Here's how to distinguish the character of a bond:

- ✔ Usually, a difference in electronegativity less than about 0.3 means that the corresponding covalent bond is considered nonpolar.

- ✔ Differences in electronegativity ranging from 0.3 to about 1.7 correspond to increasingly polar covalent bonds.

- ✔ Above about 1.7, the bond is increasingly considered ionic.

The electronegativities of the elements are displayed in Figure 5-10 and help make clear why atoms that lie horizontally farther from each other on the periodic table tend to form more polar bonds.

Differences in the electronegativity of atoms create polarity in bonds between those atoms. The polar bonds within a molecule add to create polarity in the molecule as a whole. The precise way in which the individual bonds contribute to the overall polarity of the molecule depends on the shape of the molecule.

- ✔ If two very polar bonds point in opposite directions, their polarities cancel out.

- ✔ If the two polar bonds point in the same direction, their polarities add.

- ✔ If the two polar bonds point at each other so they're diagonal to one another, their polarities cancel in one direction but add in a perpendicular direction.

Because electrons are distributed unevenly within a polar covalent bond, the more electronegative atom takes on a partial negative charge, signified by the symbol $\delta-$. The less electronegative atom takes on a partial positive charge, $\delta+$. This difference in charge along the axis of the bond is called a *dipole*. Individual bonds have dipoles, which sum over all the bonds of a molecule (taking geometry into account) to create a *molecular dipole*. In addition to the permanent dipoles created by polar bonds, instantaneous dipoles can be temporarily created within nonpolar bonds and molecules. Both kinds of dipoles play important roles in the interactions between molecules. Permanent dipoles lead to dipole-dipole interactions and to hydrogen bonds. Instantaneous dipoles lead to attractive London forces (flip to Chapters 10 and 12 for details on how these kinds of forces affect molecules in solutions).

Figure 5-10:
Electronegativities of the elements.

Decreasing →

Increasing ↑

Electronegativities of the Elements

1 H 2.1																	
3 Li 1.0	4 Be 1.5											5 B 2.0	6 C 2.5	7 N 3.0	8 O 3.5	9 F 4.0
11 Na 0.9	12 Mg 1.2											13 Al 1.5	14 Si 1.8	15 P 2.1	16 S 2.5	17 Cl 3.0
19 K 0.8	20 Ca 1.0	21 Sc 1.3	22 Ti 1.5	23 V 1.6	24 Cr 1.6	25 Mn 1.5	26 Fe 1.8	27 Co 1.9	28 Ni 1.9	29 Cu 1.9	30 Zn 1.6	31 Ga 1.6	32 Ge 1.8	33 As 2.0	34 Se 2.4	35 Br 2.8
37 Rb 0.8	38 Sr 1.0	39 Y 1.2	40 Zr 1.4	41 Nb 1.6	42 Mo 1.8	43 Tc 1.9	44 Ru 2.2	45 Rh 2.2	46 Pd 2.2	47 Ag 1.9	48 Cd 1.7	49 In 1.7	50 Sn 1.8	51 Sb 1.9	52 Te 2.1	53 I 2.5
55 Cs 0.7	56 Ba 0.9	57 La 1.1	72 Hf 1.3	73 Ta 1.5	74 W 1.7	75 Re 1.9	76 Os 2.2	77 Ir 2.2	78 Pt 2.2	79 Au 2.4	80 Hg 1.9	81 Tl 1.8	82 Pb 1.9	83 Bi 1.9	84 Po 2.0	85 At 2.2
87 Fr 0.7	88 Ra 0.9	89 Ac 1.1														

Q. Why doesn't it make sense to ask whether an element (like hydrogen or fluorine) engages in polar bonds versus nonpolar bonds?

A. Whether an element engages in polar or nonpolar bonds can only be answered with respect to specific bonds with other elements. For example, hydrogen engages in a perfectly nonpolar covalent bond with another hydrogen atom in the molecule H_2. Likewise, fluorine engages in a nonpolar bond with another fluorine atom in F_2. On the other hand, the bond between hydrogen and fluorine in the compound HF is very polar. The polarity of a bond depends on the difference in electronegativity between the bonded atoms.

10. Predict whether bonds between the following pairs of atoms are nonpolar covalent, polar covalent, or ionic:

a. H and Cl

b. Ga and Ge

c. O and O

d. Na and Cl

e. C and O

Solve It

11. Tetrafluoromethane (CF_4) contains four covalent bonds. Water (H_2O) contains two covalent bonds. Which molecule has bonds with more polar character? Which is the more polar molecule, and why?

Solve It

Shaping Molecules: VSEPR Theory and Hybridization

We'll start with the hard part: VSEPR stands for *valence shell electron pair repulsion*. Okay, now it gets easier. VSEPR is simply a model that helps predict and explain why molecules have the shapes they do. Molecule shapes help determine how molecules interact with each other. For example, molecules that stack nicely on one another are more likely to form solids. And two molecules that can fit together so their reactive bits lie closer together in space are more likely to react with one another.

The basic principle underlying VSEPR theory is that valence electron pairs, whether they are lone pairs or occur within bonds, prefer to be as far from one another as possible. There's no sense in crowding negative charges any more than necessary.

Of course, when multiple pairs of electrons participate in double or triple covalent bonds, those electrons stay within the same bonding axis. Lone pairs repel other lone pairs more strongly than they repel bonding pairs, and the weakest repulsion is between two pairs of bonding electrons. Two lone pairs separate themselves as far apart as they can go, on exact opposite sides of an atom if possible. Electrons involved in bonds also separate themselves as far apart as they can go, but with less force than two lone pairs. In general, all electron pairs try to maintain the maximum mutual separation. But when an atom is bonded to many other atoms, the "ideal" of maximum separation isn't always possible because bulky groups bump into one another. So, the final shape of a molecule emerges from a kind of negotiation between competing interests.

VSEPR theory predicts several shapes that appear over and over in real-life molecules. These shapes are summarized in Figure 5-11.

Consider one beautifully symmetrical shape predicted by VSEPR theory: the tetrahedron. Four equivalent pairs of electrons in the valence shell of an atom should distribute themselves into such a shape, with equal angles and equal distance between each pair. But what sort of atom has four *equivalent* electron pairs in its valence shell? Aren't valence electrons distributed between different kinds of orbitals, like *s* and *p* orbitals?

In order for VSEPR theory to make sense, it must be combined with another idea: *hybridization*. Hybridization refers to the mixing of atomic orbitals into new, hybrid orbitals. Electron pairs occupy equivalent hybrid orbitals. It's important to realize that the hybrid orbitals are all equivalent, because that helps you understand the shapes that emerge from the electron pairs trying to distance themselves from one another. If electrons in a pure *p* orbital are trying to distance themselves from electrons in another *p* orbital *and* from electrons in an *s* orbital, the resulting shape may not be symmetrical because *s* orbitals are different from *p* orbitals. But if all these electrons occupied identical hybrid orbitals (each orbital a little bit *s* and a little bit *p*), then the resulting shape is more likely to be symmetrical.

Real molecules have all sorts of symmetrical shapes that just don't make sense if electrons truly occupy only "pure" orbitals (like s and *p*). The "mixing" of pure orbitals into hybrids allows chemists to explain the symmetrical shapes of real molecules with VSEPR theory. This kind of mixing must in some sense actually occur, as the case of methane, CH_4, makes clear.

The shape of methane, confirmed by experiment, is tetrahedral. The four C-H bonds of methane are of equal strength and are equidistant from each other. Now compare that description with the electron configuration of carbon in Figure 5-12. Carbon contains a filled 1s orbital, but this is an inner-shell orbital, so it doesn't impact the geometry of bonding. However, the valence shell of carbon contains one filled 2s orbital, two half-filled 2p orbitals, and one empty *p* orbital (see Chapter 4 for more about these orbitals). Not the picture of equality. This configuration is inconsistent with the tetrahedral bonding geometry of carbon, making clear that the valence orbitals must hybridize.

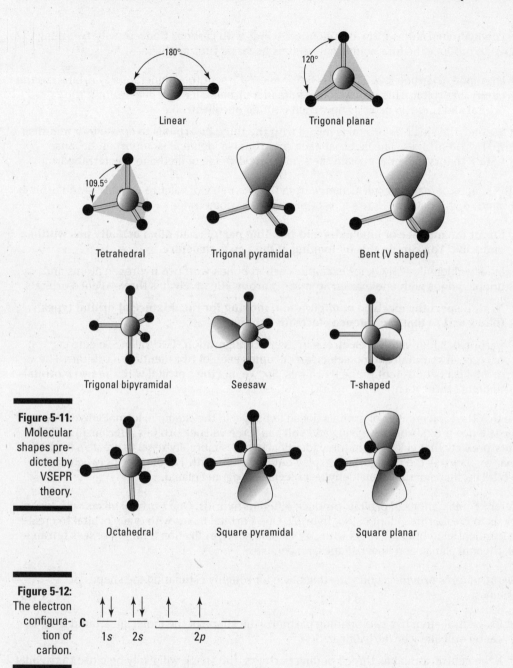

Figure 5-11:
Molecular shapes predicted by VSEPR theory.

Linear

Trigonal planar

Tetrahedral

Trigonal pyramidal

Bent (V shaped)

Trigonal bipyramidal

Seesaw

T-shaped

Octahedral

Square pyramidal

Square planar

Figure 5-12:
The electron configuration of carbon.

C $1s$ $2s$ $2p$

How can hydrogen atoms form four identical bonds with carbon? Conceptually, two things happen to produce the four equivalent orbitals necessary for methane.

- First, one of carbon's $2s$ electrons is "promoted" (sent to a higher-energy orbital) to the empty $2p$ orbital. This promotion results in four half-filled valence orbitals; the $2s$ orbital and three $2p$ orbitals now each contain one electron.

- Second, the single $2s$ orbital is mixed with the three $2p$ orbitals to create four identical sp^3 hybrid orbitals. The fact that each sp^3 orbital is identical is important because VSEPR theory can now explain the symmetrical shape of methane: the tetrahedron.

So, the shapes of real molecules emerge from the geometry of valence orbitals — the orbitals that bond to other atoms. Here's how to predict this geometry:

1. **Count the number of lone pairs and bonding partners an atom actually has within a molecule. You can do this by looking at the Lewis structure.**

In formaldehyde (CH_2O), for example, carbon bonds with two hydrogen atoms and double bonds with one oxygen atom. So, carbon effectively has three valence orbitals.

2. **Next, inspect the electron configuration, looking for the mixture of orbital types (like s and p) that the valence electrons occupy.**

Carbon has four valence electrons in $2s^2 2p^2$ configuration. Two valence electrons occupy an s orbital, and one electron occupies each of two identical p orbitals. The s orbital isn't equivalent to the p orbitals. So, we mix the s orbital with the two p orbitals to create three identical sp^2 hybrid orbitals.

Note that the total number of orbitals doesn't change; in the example, formaldehyde has three valence orbitals before mixing and still has three valence orbitals after mixing. VSEPR theory predicts that electrons in three identical orbitals mutually repel to create a *trigonal planar geometry* — the three orbitals splay out in a plane with 120 degrees between each orbital. The shape of the formaldehyde molecule is trigonal planar.

Different combinations of orbitals produce different hybrids. One s orbital mixes with two p orbitals to create three identical sp^2 hybrids. One s orbital mixes with one p orbital to create two identical sp hybrids. Centers with sp^3, sp^2, and sp hybridization tend to possess tetrahedral, trigonal planar, and linear shapes, respectively.

Lewis structures provide a good starting place for roughly estimating the shapes of molecules.

- If a central atom has two bonding partners, the shape of the molecule around that center will likely be fairly linear.

- If the central atom has three bonding partners, the shape will likely be close to trigonal planar.

- If the central atom has four bonding partners, the shape will likely be close to tetrahedral.

The actual shapes may vary from these rough estimates depending on other factors, such as whether the central atom has lone pairs. For example, the oxygen atom of water has two lone pairs in addition to two bonding orbitals. In all, then, the electrons in the orbitals surrounding oxygen create a tetrahedral-like *geometry*. But, because only two hydrogen atoms are actually bound, the shape of the *molecule* is "bent."

EXAMPLE

Q. Methane, CH_4, has four hydrogen atoms bonded to a central carbon atom. Ammonia, NH_3, has three hydrogen atoms bonded to a central nitrogen atom. Using VSEPR theory, compare and contrast the *orbital geometry* and *molecular shape* of these two molecules.

A. Carbon and nitrogen are both sp^3 hybridized in methane and ammonia, respectively. Carbon contributes a single electron in each hybrid orbital to a covalent bond, the second electron in each bond coming from a hydrogen. Nitrogen has one more valence electron than carbon. Therefore, one of the hybrid orbitals of nitrogen contains two electrons and can't receive a bonding electron from hydrogen. Although ammonia has only three bonded hydrogen atoms for this reason, the central nitrogen has a lone pair of electrons in the remaining sp^3 orbital. These electrons still repel other electron pairs. So, the ammonia molecule has a nearly tetrahedral orbital geometry, similar to that of methane. However, because one of the orbitals contains a lone pair (not a bonded atom) the molecular shape of ammonia is trigonal pyramidal.

---

12. What's the hybridization of carbon in carbon dioxide (CO_2)? In formaldehyde (CH_2O)? In methyl bromide (H_3CBr)?

Solve It

13. Use Lewis structures and VSEPR theory to predict the shape of water (H_2O), ethyne (C_2H_2), and carbon tetrachloride (CCl_4).

Solve It

---

14. Chlorine trifluoride (ClF_3) is a T-shaped molecule with trigonal bipyramidal orbital geometry. How is the T shape possible?

Solve It

Answers to Questions on Bonds

See how well you've bonded to the concepts in this chapter. Check your answers to find out if they overlap with the ones provided here. If not, take another orbit through the questions.

1 Potassium (K) transfers its single valence electron to fluorine (F), yielding an ionic bond between K^+ and F^-, as shown in the following figure:

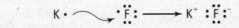

2 **Two lithium atoms each transfer a single electron to one sulfur atom** to yield the ionic compound, Li_2S. As a group IA metal, lithium easily gives up its single valence electron. As a group VIA nonmetal, sulfur readily accepts two additional electrons into its valence shell.

3 As a salt, magnesium chloride ($MgCl_2$) is certainly an ionic compound. So, $MgCl_2$ dissolves to a greater extent in more polar solvents. Dissolved ions act as *electrolytes,* conducting electricity in solutions. The more brightly glowing bulb in the circuit containing the aqueous (water-based) solution suggests that more electrolytes are present in that solution. More salt dissolved in water than in rubbing alcohol because water is the more polar solvent.

4 Dichlorine is Cl_2, a compound formed when one chlorine atom bonds to another. Because each atom in the compound is of the same element, the two atoms have the same electronegativity. So, the difference in electronegativity between the two atoms is 0. This means that the bond between the two chlorine atoms must be covalent. The electron dot structure of dichlorine is shown in the following figure:

$$:\ddot{\text{Cl}}:\ddot{\text{Cl}}:$$

5 A **coordinate covalent bond** forms between the aluminum and the hydrating water molecules. Aluminum is a group IIIA element, so the aluminum (III) cation formally has no valence electrons. The oxygen of water has lone pairs. So, water molecules most likely hydrate the cation by donating lone pairs to form coordinate covalent bonds. In this respect, we can call the water molecules ligands of the metal ion.

6 Resonance structures for benzene are shown in the following figure. Adjacent carbons in the ring are held together with either single or double covalent bonds depending on the resonance structure. So, in the resonance hybrid structure, each carbon-carbon bond is identical, and is neither a single bond nor a double bond, but something more like a one-and-a-half bond.

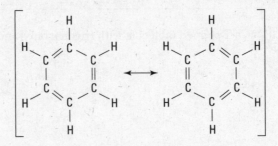

7 The molecular orbital diagram for dihelium is shown in the following figure. Each helium atom contributes two electrons to the molecular orbitals for a total of four electrons. Each molecular orbital holds two electrons, so both the low-energy bonding orbital and the high-energy antibonding orbitals are filled.

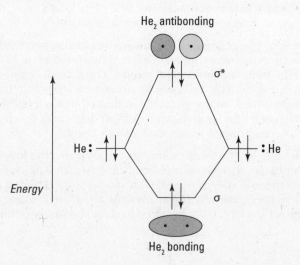

He$_2$ antibonding

σ^*

He

: He

Energy

σ

He$_2$ bonding

8 The total energy change to make dihelium from two separate helium atoms is the sum of the changes due to bonding and antibonding. Putting two electrons in the antibonding orbital costs more energy than is saved by putting two electrons in the bonding orbital. So, going from two separate helium atoms to one molecule of dihelium requires an input of energy. Reactions spontaneously go to lower energy, not higher energy, so dihelium is unlikely to exist under normal conditions.

9 The reaction breaks one of the carbon-carbon bonds, replacing the double bond with two single bonds. The carbon-carbon bond that breaks is the pi bond, because the pi electrons are more accessible (above and below the axis of the bond) and because the pi bond is weaker than the sigma bond. So, the weaker pi bond is replaced by a stronger sigma bond. In bonds, *weaker* means higher energy, and *stronger* means lower energy. So, the reaction moves from higher to lower energy, which is favorable.

10 The assignment of bond character depends on the difference in electronegativity between the atoms:

a. The difference in electronegativity between H and Cl is 0.9, so the bond is a **polar covalent bond.**

b. The difference in electronegativity between Ga and Ge is 0.2, so the bond is a **nonpolar covalent bond.**

c. The difference in electronegativity between O and O is 0.0, so the bond is a **nonpolar covalent bond.**

d. The difference in electronegativity between Na and Cl is 2.1, so the bond is an **ionic bond.**

e. The difference in electronegativity between C and O is 1.0, so the bond is a **polar covalent bond.**

11 Both molecules contain polar covalent bonds. The C-F bonds of CF_4 are slightly more polar than the H-O bonds of water due to the greater difference in electronegativity between the former pair of atoms (1.5 versus 1.4). However, water is a much more polar molecule due to molecular shape. Because of its tetrahedral shape, the bond dipoles of CF_4 cancel each other out entirely, so the molecule as a whole is nonpolar. Because of its bent shape, the bond dipoles of water only partially cancel, so the molecule as a whole is polar.

12 **In CO_2, carbon is *sp* hybridized.** CO_2 is linear, with the carbon double-bonded to each oxygen so that four electrons are constrained in each of two bond axes. **In CH_2O, carbon is *sp^2* hybridized, so the molecule is trigonal planar.** The four electrons of the double bond are constrained within one bond axis, and two electrons each are within bonds to hydrogen. **In H_3CBr, carbon is *sp^3* hybridized, so the molecule is shaped like a tetrahedron, similar to the shape of methane, CH_4;** each bond is separated from all the others by approximately 109 degrees. Two electrons are constrained within each of four bonds.

13 The Lewis structures for the three molecules are shown in the following figure. The oxygen of water is sp^3 hybridized, with two single bonds and two lone pairs, resulting in a bent shape. Each carbon of ethyne is sp hybridized, with six electrons devoted to a triple bond and two devoted to a single bond with hydrogen, resulting in a linear shape. Carbon tetrachloride is sp^3 hybridized, with all electron pairs involved in single bonds, resulting in a tetrahedral shape.

$$H-C\equiv C-H$$

$$
\begin{array}{c}
Cl \\
| \\
Cl - C - Cl \\
| \\
Cl
\end{array}
$$

14 The hybridization of an atom's valence orbitals determines that atom's orbital geometry, but doesn't completely determine molecular shape. Shape depends on whether electrons in the orbitals are involved in bonding or in lone pairs. Trigonal bipyramidal geometry suggests that there are five valence orbitals: two along a central axis and three distributed evenly in a plane perpendicular to that axis (it may help to refer back to Figure 5-10). In ClF_3, Cl is the central atom. Both orbitals along the central axis are involved in Cl-F bonds to fluorine atoms. Only one of the orbitals in the perpendicular plane is involved in a bond to the remaining fluorine atom. The other two orbitals contain lone pairs. The result is a T-shaped molecule.

Chapter 6

Naming Compounds

● ●

In This Chapter

▶ Crafting names for ionic and molecular compounds

▶ Handling polyatomic ions

▶ Using a quick and easy scheme for naming any compound

● ●

Chemists give compounds very specific names. Sometimes these names seem overly specific. For example, what's the point of referring to "dihydrogen monoxide" when you can simply say "water"? First, you must try to understand that many chemists simply believe this kind of thing sounds cool. Beyond dubious notions of coolness, however, there lies a more important reason: Chemical names clue you in to chemical structures. But only if you know the code. Fortunately, the code, as you find out in this chapter, is pretty straightforward — no advanced cryptology required. This is also fortunate, because putting chemists and cryptologists in the same room could result in the kind of party you don't want to admit having attended.

Naming Ionic Compounds

Chapter 5 discusses the way that *anions* (atoms with negative charge) and *cations* (atoms with positive charge) attract one another to form ionic bonds. Ionic compounds are held together by ionic bonds. Fine, but how does the *formula* of an ionic compound relate to the *name* of the compound?

Naming a simple ionic compound is easy. You pair the name of the cation with the name of the anion, and then change the ending of the anion's name to *–ide*. The cation always precedes the anion in the final name. For example, the chemical name of NaCl (a compound made up of one sodium atom and one chlorine atom) is sodium chlor*ide*.

Of course, sodium chloride is more commonly known as table salt. Many compounds have such so-called *common names*. There isn't anything necessarily wrong with common names, but they're less informative than chemical names. The name "sodium chloride," properly decoded, tells you that you're dealing with a one-to-one ionic compound composed of sodium and chlorine.

Cations and anions combine in very predictable ways within ionic compounds, always acting to neutralize overall charge. Because this is true, the name of an ionic compound implies more than just the identity of the atoms that make it up. It also implies the ratio in which elements combine. Consider these two examples, both of which involve lithium:

✔ In the compound lithium fluoride (LiF), lithium and fluorine combine in a one-to-one ratio because lithium's +1 charge and fluorine's –1 charge cancel one another (neutralize) perfectly. By itself, the name lithium fluoride tells us only that the compound is made up of a lithium cation and a fluoride anion, but by comparing their charges, you can see that Li^+ and F^- neutralize each other in the one-to-one compound, LiF.

✔ If lithium combines with oxygen to form an ionic compound, two lithium ions, each with +1 charge, are required to neutralize the –2 charge of the oxide anion. So, the name lithium oxide implies the formula Li_2O, because a two-to-one ratio of lithium cation to oxide anion is required to produce a neutral compound.

Use the name of the ionic compound to identify the ions you're dealing with, and then combine those ions in the simplest way that results in a neutral compound.

Using a compound's name to identify the ions can be tricky when the cation is a metal. All group B metals — with the exception of silver (which is always found as Ag^+) and zinc (always Zn^{2+}) — as well as several group A elements on the right-hand side of the periodic table can take on a variety of charges. Chemists use an additional naming device, the Roman numeral, to identify the charge state of the cation. A Roman numeral placed within parentheses after the name of the cation gives the positive charge of that cation. For example, copper (I) is copper with a +1 charge, and copper (II) is copper with a +2 charge. Is this kind of distinction simply chemical hair-splitting? Who cares whether you're dealing with iron (II) bromide, $FeBr_2$, or iron (III) bromide, $FeBr_3$? The difference matters because ionic compounds with different formulas (even those containing the same types of elements) can have very different properties.

The trickiness of metals with variable charges also applies when you translate a chemical formula into the corresponding chemical name. Here are the two secrets to keep in mind:

✔ The name implies the charge of the ions which make up an ionic compound.

✔ The charges of the ions determine the ratio in which they combine.

Given the formula CrO, for example, you must do a little bit of sleuthing before assigning a name. From periodic table trends (which we cover in Chapter 4), you know that oxygen brings a –2 charge to an ionic compound. Because O^{2-} combines with chromium in a one-to-one ratio within CrO, you know that chromium here must have the equal and opposite charge of +2. In this way, CrO can be electrically neutral (uncharged). So, the chromium cation in CrO is Cr^{2+}, and the compound name is chromium (II) oxide. Simply calling the compound chromium oxide leaves open the question of the precise chromium cation in play. For example, chromium can assume a +3 charge. When Cr^{3+} combines with oxide anion, a different ion ratio is necessary to produce a neutral compound: Cr_2O_3, or chromium (III) oxide. This equating of total positive and negative charges in an atom is called *balancing the charge*. After you have balanced the charges in the atom's formula, you can drop the charges on the individual ions.

EXAMPLE

Q. What is the formula for the compound tin (IV) fluoride?

A. **SnF$_4$.** The Roman numeral within parentheses tells you that you're dealing with Sn^{4+}. Because fluorine is a halogen, it always has a charge of –1 in ionic compounds, which means that four fluoride anions are necessary to cancel the four positive charges of a single tin cation. Therefore, the compound is SnF$_4$.

---

1. Name the following compounds:

a. MgF$_2$

b. LiBr

c. Cs$_2$O

d. CaS

Solve It

2. Name the following compounds that contain elements with variable charge. Don't forget to use Roman numerals!

a. FeF$_2$

b. HgBr

c. SnI$_4$

d. Mn$_2$O$_3$

Solve It

---

3. Translate the following names into chemical formulas:

a. Iron (III) oxide

b. Beryllium chloride

c. Tin (II) sulfide

d. Potassium iodide

Solve It

Dealing with Those Pesky Polyatomic Ions

A confession: We have shielded you thus far from a very disturbing fact about ions. Not all ions are of single atoms. To continue your education in chemical nomenclature, you must grapple with an irksome and all-too-common group of molecules called the *polyatomic ions,* which are made up of groups of atoms. Polyatomic ions, like single-element ions, tend to quickly combine with other ions to neutralize their charge. Unfortunately, you can't use any simple, periodic trend–type rules to figure out the charge of a polyatomic ion. You must — gulp — memorize them. Truth.

Table 6-1 summarizes the most common polyatomic ions, grouping them by charge.

Table 6-1	Common Polyatomic Ions
−1 Charge	*−2 Charge*
Dihydrogen phosphate ($H_2PO_4^-$)	Hydrogen phosphate (HPO_4^{2-})
Acetate ($C_2H_3O_2^-$)	Oxalate ($C_2O_4^{2-}$)
Hydrogen sulfite (HSO_3^-)	Sulfite (SO_3^{2-})
Hydrogen sulfate (HSO_4^-)	Sulfate (SO_4^{2-})
Hydrogen carbonate (HCO_3^-)	Carbonate (CO_3^{2-})
Nitrite (NO_2^-)	Chromate (CrO_4^{2-})
Nitrate (NO_3^-)	Dichromate ($Cr_2O_7^{2-}$)
Cyanide (CN^-)	Silicate (SiO_3^{2-})
Hydroxide (OH^-)	*−3 Charge*
Permanganate (MnO_4^-)	Phosphite (PO_3^{3-})
Hypochlorite (ClO^-)	Phosphate (PO_4^{3-})
Chlorite (ClO_2^-)	*+1 Charge*
Chlorate (ClO_3^-)	Ammonium (NH_4^+)
Perchlorate (ClO_4^-)	

Notice in Table 6-1 that all the common polyatomic ions, except ammonium, have a negative charge ranging between −1 and −3. Also take note of a number of *–ite/–ate* pairs. For example, there are chlorite and chlorate, phosphite and phosphate, and nitrite and nitrate. If you look closely at these pairs, you notice that the only difference between them is the number of oxygen atoms in each ion. Specifically, the *–ate* ion always has one more oxygen atom than the *–ite* ion, but has the same overall charge.

To complicate your life further, polyatomic ions sometimes occur multiple times within the same ionic compound. How do you specify that your compound has two sulfate ions in a way that makes visual sense? Put the entire polyatomic ion formula in parentheses, and then add a subscript outside the parentheses to indicate how many such ions you have, as in $(SO_4^{2-})_2$.

When you write a chemical formula that involves polyatomic ions, you treat them just like other ions. You still need to balance charges to form a neutral atom. When converting from a formula to a name, we are sorry to report that there is no simple rule for naming polyatomic ions. You just have to memorize the entire table of polyatomic ions and their charges. Challenge a chemist to a polyatomic ion duel, and you'll find that she can rattle off the name, formula, and charge of the polyatomic ions in the blink of an eye. By the time you're done with this book, you should be able to give her a run for her money.

EXAMPLE

Q. Write the formula for the compound barium chlorite.

A. **Ba(ClO₂)₂.** Barium is an alkaline earth metal (group IIA) and thus has a charge of +2. You should recognize chlorite as the name of a polyatomic ion. In fact, any anion name which does not end in an –*ide* should scream polyatomic ion to you. As shown in Table 6-1, chlorite is ClO_2^-, which reveals that the chlorite ion has a –1 charge, so two chlorite ions are needed to neutralize the +2 charge of a single barium cation, so the chemical formula is $Ba(ClO_2)_2$.

4. Name the following compounds that contain polyatomic ions:

a. $Mg_3(PO_4)_2$

b. $Pb(C_2H_3O_2)_2$

c. $Cr(NO_2)_3$

d. $(NH_4)_2C_2O_4$

e. $KMnO_4$

Solve It

5. Write the formula for the following compounds that contain polyatomic ions:

a. Potassium sulfate

b. Lead (II) dichromate

c. Ammonium chloride

d. Sodium hydroxide

e. Chromium (III) carbonate

Solve It

Giving Monikers to Molecular Compounds

As described in Chapter 5, nonmetals tend to form covalent bonds with one another. Compounds made up of nonmetals held together by one or more covalent bonds are called *molecular compounds*. Predicting how the atoms within molecules will bond with one another is a tricky endeavor because two nonmetals often can combine in multiple ratios. Carbon and oxygen, for example, can combine in a one-to-two ratio to form CO_2 (carbon dioxide), a harmless gas that you emit every time you exhale. Alternately, the same two elements can combine in a one-to-one ratio to form CO (carbon monoxide), a poisonous gas. Clearly, it's useful to have names that distinguish between these and other molecular compounds. The punishment for sloppy naming can be death. Or at least embarrassment.

Molecular compound names clearly specify how many of each type of atom participate in the compound. The prefixes used to do so are listed in Table 6-2.

Table 6-2	Prefixes for Binary Molecular Compounds		
Prefix	*Number of Atoms*	*Prefix*	*Number of Atoms*
Mono–	1	Hexa–	6
Di–	2	Hepta–	7
Tri–	3	Octa–	8
Tetra–	4	Nona–	9
Penta–	5	Deca–	10

The prefixes in Table 6-2 can be attached to any of the elements in a molecular compound, as exemplified by SO_3 (sulfur trioxide) or N_2O (dinitrogen monoxide). The second element in each compound receives the *–ide* suffix, as in ionic compounds (which we discuss earlier in this chapter). In the case of molecular compounds, where cations or anions aren't involved, the more electronegative element (in other words, the element that's closer to the upper right-hand corner of the periodic table) tends to be named second.

Notice (with annoyance) that the absence of a prefix from the first named element of a molecular compound implies that there is only one atom of that element. In other words, the prefix *mono–* is unnecessary *for the first element only.* You still have to attach a *mono–* prefix, when appropriate, to the names of subsequent elements.

Writing a formula for a molecular compound with a given name is much simpler than writing a formula for an ionic compound. In a molecular compound, the ratio in which the two elements combine is built into the name itself, and you don't need to worry about balancing charges. For example, the prefixes in the name dihydrogen monoxide imply that the chemical formula contains two hydrogens and one oxygen (H_2O).

Translating a formula into a name is equally simple. All you need to do is convert the numbers into prefixes and attach them to the names of the elements that make up the compound. For example, for the compound N_2O_4, you simply attach the prefix *di–* to nitrogen to indicate the two nitrogen atoms, and *tetra–* to oxygen to indicate the four oxygen atoms, giving you dinitrogen tetroxide.

Hydrogen is located on the far left of the periodic table, but it's actually a nonmetal! In keeping with this hydrogenic schizophrenia, hydrogen can appear as either the first or second element in a *binary* (two-element) molecular compound, as shown by dihydrogen monosulfide (H_2S) and phosphorus trihydride, PH_3.

Q. What are the names of the compounds N_2O, SF_6, and Cl_2O_8?

A. **Dinitrogen monoxide, sulfur hexafluoride, and dichlorine octoxide.** Notice that none of these compounds contain any metals, which means that they are most certainly molecular compounds. The first compound contains two nitrogen atoms and one oxygen atom, so it's called dinitrogen monoxide. The second compound contains one sulfur and six fluorines. Because sulfur is the first named element, we needn't include a *mono–* prefix. So, we name the compound simply sulfur hexafluoride (instead of *mono*sulfur hexafluoride). Using the same methods, the third compound is named dichlorine octoxide.

6. Write the proper names for the following compounds:

a. N_2H_4

b. H_2S

c. NO

d. CBr_4

Solve It

7. Write the proper formulas for the following compounds:

a. Silicon difluoride

b. Nitrogen trifluoride

c. Disulfur decafluoride

d. Diphosphorus trichloride

Solve It

Seeing the Forest: A Unified Scheme for Naming Compounds

Have all the naming rules in the previous sections left you confused and frustrated? Do you feel like rebelling against the fascist chemical conspiracy, and liberating all chemicals from the oppressive confines of their formal names? All right, calm down. If you take a step back to look at the big picture, you see that the naming system is actually pretty logical and straightforward. You can best see the logical structure by envisioning the naming process as a flowchart, a series of questions whose answers lead down a path toward a compound's name. For any binary (two-part) compound XY, use the following four questions to guide yourself to a name.

1. **Is X hydrogen?**

Compounds which contain a hydrogen cation are commonly called *acids,* and many of them have common names. Although it is perfectly acceptable to call the compound HCl "hydrogen monochloride," it's more commonly referred to as "hydrochloric acid." As with the polyatomic ions that we cover earlier in this chapter, you'll be happier in the end if you simply memorize the common acids listed in Table 6-3.

Table 6-3	Common Acids
Name	*Formula*
Acetic acid	$H_4C_2O_2$
Carbonic acid	H_2CO_3
Hydrochloric acid	HCl
Nitric acid	HNO_3
Phosphoric acid	H_3PO_4
Sulfuric acid	H_2SO_4

2. **Is X a nonmetal or a metal?**

If X is a nonmetal, you need to use prefixes when naming both it and its partner, Y, because the compound is molecular. If X is a metal, then you're dealing with an ionic compound — proceed to Step 3.

3. **Is X a group B metal?**

If X is a group B metal (or other metal of variable charge, like tin), you need to use Roman numerals to specify its charge.

4. **Is Y a polyatomic ion?**

If Y is a polyatomic ion, you have to recognize it as such and have its name memorized (or easily accessible in a nifty table such as Table 6-1). If Y is a simple anion, then you use an *–ide* ending.

Visualize this series of questions with the help of Figure 6-1.

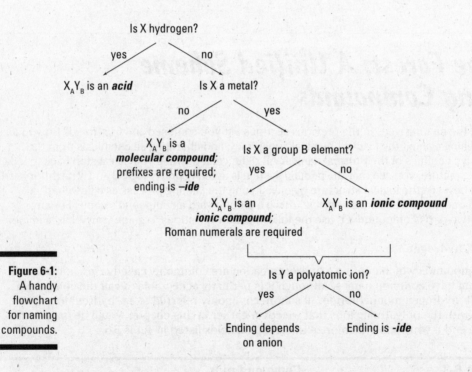

Figure 6-1:
A handy
flowchart
for naming
compounds.

You can use the flowchart in Figure 6-1 in reverse to help you figure out a chemical formula from the name of a compound. If the compound name contains a Roman numeral (III, for example), you know the charge of the metal in that compound (3+ in this case). If X is a nonmetal, you know you're dealing with a molecular compound, and you can start examining the name for prefixes that tell you how many atoms of each element there are. If you find no such prefixes, check to see whether X is hydrogen — if so, you're dealing with an acid.

EXAMPLE

Q. Name the compound $Sn(OH)_2$.

A. **Tin (II) hydroxide.** Sn is not hydrogen or a nonmetal, so you know that the compound is neither an acid nor a molecular compound. This means you're dealing with an ionic compound. Sn is, however, one of the group A elements that takes on varying charge, so you should be prepared to do some sleuthing to determine what exactly its charge is. The only remaining question is whether we have a polyatomic anion or an ordinary one. The fact that OH contains more than one atom should scream *poly*atomic ion to you. Consulting Table 6-1, you find that OH (hydroxide) has a charge of –1. If two hydroxide ions (each with a charge of –1) are needed to cancel out the positive charge of the tin, then you must be dealing with Sn^{2+}. The name of the compound is therefore tin (II) hydroxide.

8. Name each of the following compounds:

a. $PbCrO_4$

b. Mg_3P_2

c. $SrSiO_3$

d. H_2SO_4

e. Na_2S

f. B_3Se_2

g. HgF_2

h. $Ba_3(PO_4)_2$

Solve It

9. Translate the following names into chemical formulas:

a. Barium hydroxide

b. Tin (IV) bromide

c. Sodium sulfate

d. Phosphorus triiodide

e. Magnesium permanganate

f. Acetic acid

g. Nitrogen dihydride

h. Iron (II) chromate

Solve It

Answers to Questions on Naming Compounds

You survived all the practice questions on naming compounds. Now enjoy the process of seeing how spot-on correct your answers were.

1. These are all ordinary ionic compounds, so you simply need to pair the cation name with the anion name and change the anion name's ending to –*ide*.

   a. **Magnesium fluoride**

   b. **Lithium bromide**

   c. **Cesium oxide**

   d. **Calcium sulfide**

2. As the problem states, all the cations here are ones that can have varying amounts of positive charge, so you need to decipher their charges.

   a. **Iron (II) fluoride.** The fluoride ion has a charge of –1. Because there are two fluorides here, the single iron ion must have a +2 charge.

   b. **Mercury (I) bromide.** The –1 charge of the bromide must be balanced by a +1 charge.

   c. **Tin (IV) iodide.** The four iodide anions each have a –1 charge. This means that the tin cation must have a charge of +4.

   d. **Manganese (III) oxide.** There are three oxide anions here, each with a charge of –2, giving an overall negative charge of –6. The manganese cations must carry a total of +6 charge, split between the two cations, so we must be dealing with Mn^{3+}.

3. The names translate into the following chemical formulas:

   a. **Fe_2O_3.** Because the name specifies that you're dealing with Fe^{3+}, and because oxygen is always O^{2-}, you simply balance your charges to yield Fe_2O_3.

   b. **$BeCl_2$.** Beryllium is an alkaline earth metal with a charge of +2, while chlorine is a halogen with a charge of –1.

   c. **SnS.** Because the name specifies that you are dealing with Sn^{2+}, and because sulfur is always S^{2-}, you simply balance your charges to yield SnS.

   d. **KI.** Potassium is an alkali metal with a charge of +1, while iodine is a halogen with a charge of –1.

4. Look up (or recall) the polyatomic ions in each compound, and specify the charge of the cation if it's a metal that can take on different charges.

   a. **Magnesium phosphate.** Magnesium is the cation here, and the anion is PO_4. Table 6-1 tells you that this is the polyatomic ion phosphate (not to be confused with phosphite, or PO_3).

   b. **Lead (II) acetate.** Acetate has a –1 charge, and because two of them are needed to balance out the charge on the lead cation, you must be dealing with Pb^{2+}.

   c. **Chromium (III) nitrite.** Nitrite has a –1 charge, and because three of them are needed to balance out the charge on the chromium cation, you must be dealing with Cr^{3+}.

> **d. Ammonium oxalate.** Here, both the cation and the anion are polyatomic ions. Annoying but true.
>
> **e. Potassium permaganate.** Potassium is the cation here, so all you need to do is look up the anion MnO_4 to find that it's called permaganate.

5 First look up (or better, recall from memory) the charge of the polyatomic ion or ions, and then use subscripts as necessary to balance charges.

  a. K_2SO_4

  b. $PbCr_2O_7$

  c. NH_4Cl

  d. $NaOH$

  e. $Cr_2(CO_2)_3$

6 Translate the subscripts into prefixes using Table 6-2. Omit the prefix *mono–* on the first named element in a compound, where applicable.

  a. Dinitrogen tetrahydride

  b. Dihydrogen monosulfide

  c. Nitrogen monoxide

  d. Carbon tetrabromide

7 Translate the prefixes into subscripts using Table 6-2. If the first named element lacks a prefix, assume that there is only one such atom per molecule.

  a. SiF_2

  b. NF_3

  c. S_2F_{10}

  d. P_2Cl_3

8 Use the flowchart in Figure 6-1 to guide yourself to a name.

  a. Lead (II) chromate. Lead is a group B element, and CrO_4^{2-} is a polyatomic ion. Therefore, you need to determine the charge on the lead and specify that charge with a Roman numeral. If you haven't memorized it, find CrO_4^{2-} in Table 6-1. Because chromate combines with lead in a one-to-one ratio, you know that the charge on the lead must be +2.

  b. Magnesium phosphide. This is a simple ionic compound, because Mg is a non-group B metal, and because P is not a polyatomic ion.

  c. Strontium silicate. Sr is neither a group B element nor a nonmetal, but SiO_3 forms a polyatomic ion.

  d. Sulfuric acid (common name) or **dihydrogen sulfate** (systematic name). In addition, this compound is sometimes referred to simply as hydrogen sulfate because the –2 charge on sulfate implies that two hydrogen cations are necessary to neutralize charge. The cation here is hydrogen, so you're dealing with an acid. Common names of acids are listed in Table 6-3. You can also name the compound properly by recognizing the polyatomic ion, sulfate.

   e. Sodium sulfide. Simply name the cation and change the ending of the (polyatomic) anion to *–ide*.

   f. Triboron diselenide. Both boron and selenium are nonmetals, so this is a molecular compound. Therefore, it must be named using prefixes.

   g. Mercury (II) fluoride. Here you have an ionic compound with a group B metal, necessitating the use of a Roman numeral. The charge on the mercury atom must be +2 because it combines with two fluoride anions, each of which must have a –1 charge.

   h. Barium phosphate. This is an ionic compound which contains a polyatomic ion, phosphate.

9 Reverse your naming rules to deduce the chemical formula of each compound.

   a. $Ba(OH)_2$. Barium is an ordinary metal, while hydroxide is a polyatomic ion with a charge of –1.

   b. $SnBr_4$. The name indicates that you are dealing with Sn^{4+}, and the fluoride ion has a charge of –1, so you need four of them to balance the charge of a single tin cation.

   c. Na_2SO_4. Sodium is a simple alkali metal, while sulfate is a polyatomic ion with a charge of –2.

   d. PI_3. The prefixes indicate that this is a molecular compound containing a single phosphorus atom and three iodine atoms.

   e. $Mg(MnO_4)_2$. Magnesium is a metal, while permanganate is a polyatomic ion with a charge of –1.

   f. $H_4C_2O_2$. Use Table 6-3 to translate this common acid into its chemical formula.

   g. NH_2. The *di–* prefix and the fact that both elements are nonmetals indicate that this is a molecular compound. Nitrogen, the first named element, lacks a prefix, so there must be only one nitrogen per molecule. The *di–* prefix indicates two hydrogen atoms per molecule.

   h. $FeCrO_4$. The name tells you that you're dealing with Fe^{2+} and the polyatomic ion chromate, which has a charge of –2. A one-to-one ratio is sufficient to neutralize overall charge.

Chapter 7

Managing the Mighty Mole

. . . .

In This Chapter

▶ Making particle numbers manageable with Avogadro's number

▶ Converting between masses, mole counts, and volumes

▶ Dissecting compounds with percent composition

▶ Moving from percent composition to empirical and molecular formulas

. . . .

Chemists routinely deal with hunks of material containing trillions of trillions of atoms, but ridiculously large numbers can induce migraines. For this reason, chemists count particles (like atoms and molecules) in multiples of a quantity called the *mole*. Initially, counting particles in moles can be counterintuitive, similar to the weirdness of Dustin Hoffman's character in the movie *Rainman,* when he informs Tom Cruise's character that he has spilled 0.41 fraction of a box of 200 toothpicks onto the floor. Instead of referring to 82 individual toothpick particles, he refers to a fraction of a larger unit, the 200-toothpick box. A mole is a very big box of toothpicks — 6.022×10^{23} toothpicks, to be precise. In this chapter, we explain what you need to know about moles.

Counting Your Particles: The Mole

If 6.022×10^{23} strikes you as an unfathomably large number, then you're thinking about it correctly. It's larger, in fact, than the number of stars in the sky or the number of fish in the sea, and is many, many times more than the number of people who have ever been born throughout all of human history. When you think about the number of particles in something as simple as, say, a cup of water, all your previous conceptions of "big numbers" are blown out of the water, as it were.

The number 6.022×10^{23}, known as *Avogadro's number,* is named after the 19th century Italian scientist Amedeo Avogadro. Posthumously, Avogadro really pulled one off in giving his name to this number, because he never actually thought of it. The real brain behind Avogadro's number was that of a French scientist named Jean Baptiste Perrin. Nearly 100 years after Avogadro had his final pasta, Perrin named the number after him as an homage. Ironically, this humble act of tribute has misdirected the resentment of countless hordes of high school chemistry students to Avogadro instead of Perrin.

Avogadro's number is the conversion factor used to move between particle counts and numbers of moles:

$$\frac{1 \text{ mol}}{6.022 \times 10^{23} \text{ particles}}$$

Like all conversion factors, you can invert it to move in the other direction, from moles to particles. (Flip to Chapter 2 for an introduction to conversion factors.)

Q. How many water molecules are in one tablespoon of water, if the tablespoon holds 0.82 moles?

A. **4.9 × 10²³ molecules.** To convert from moles to particles, you need a conversion factor with moles in the denominator and particles in the numerator.

$$\frac{0.82 \text{ mol}}{1} \times \frac{6.022 \times 10^{23} \text{ molecules}}{1 \text{ mol}} = 4.9 \times 10^{23} \text{ molecules}$$

The units of moles in the numerator of the first term and the denominator of the second term cancel, so you're left with the final answer of 4.9×10^{23} molecules.

1. If the average galaxy has approximately 150 billion stars, and if astronomers estimate that the universe has at least 125 billion galaxies, how many moles of stars are in the universe?

2. On Earth, there are roughly 200 million insects for every single human being. If the current world population is nearly 7 billion, how many moles of bugs are on the planet?

Solve It

3. How many atoms of carbon-14 are in the average human body given the following four facts:

1. Carbon makes up 23% of the human body by weight.

2. The average human body has a mass of 150 kg.

3. There are approximately 83 moles in every kg of carbon.

4. One in every trillion atoms of carbon is carbon-14.

Solve It

Assigning Mass and Volume to Moles

Chemists always begin a discussion about moles with Avogadro's number. They do this for two reasons. First, it makes sense to start the discussion with the way the mole was originally defined. Second, it's a sufficiently large number to intimidate the unworthy.

Still, for all its primacy and intimidating size, Avogadro's number quickly grows tedious in everyday use. More interesting is the fact that one mole of a pure *monatomic* substance (in other words, a substance that always appears as a single atom) turns out to possess exactly its atomic mass's worth of grams. In other words, one mole of monatomic hydrogen weighs about 1 gram. One mole of monatomic helium weighs about 4 grams. The same is true no matter where you wander through the corridors of the periodic table. The number listed as the atomic mass of an element equals that element's *molar mass* if the element is monatomic.

Of course, chemistry involves the making and breaking of bonds (as you find out in Chapter 5), so talk of pure monatomic substances gets you only so far. How lucky, then, that calculating the mass per mole of a complex molecule is essentially no different than finding the mass per mole of a monatomic element. For example, one molecule of glucose ($C_6H_{12}O_6$) is assembled from 6 carbon atoms, 12 hydrogen atoms, and 6 oxygen atoms. To calculate the number of grams per mole of a complex molecule (such as glucose), simply do the following:

1. **Multiply the number of atoms per mole of the first element by its atomic mass.**

In this case, the first element is carbon, and you'd multiply its atomic mass, 12, by the number of atoms, 6.

2. **Multiply the number of atoms per mole of the second element by its atomic mass.**

Here, you multiply hydrogen's atomic mass of 1 by the number of hydrogen atoms, 12.

3. **Multiply the number of atoms per mole of the third element by its atomic mass. Keep going until you've covered all the elements in the molecule.**

The third element in glucose is oxygen, so you multiply 16, the atomic mass, by 6, the number of oxygen atoms.

4. **Finally, add the masses together.**

In this example, $(12g\ mol^{-1} \times 6) + (1g\ mol^{-1} \times 12) + (16g\ mol^{-1} \times 6) = 180g\ mol^{-1}$.

This kind of quantity, called the *gram molecular mass,* is exceptionally convenient for chemists, who are much more inclined to measure the mass of a substance than to count all of the millions or billions of individual particles that make it up.

If chemists don't try to intimidate you with large numbers, they may attempt to do so by throwing around big words. For example, chemists may distinguish between the molar masses of pure elements, molecular compounds, and ionic compounds by referring to them as the *gram atomic mass, gram molecular mass,* and *gram formula mass,* respectively. Don't be fooled! The basic concept behind each is the same: molar mass.

It's all very good to find the mass of a solid or liquid and then go about calculating the number of moles in that sample. But what about gases? Let's not engage in phase discrimination; gases are made of matter, too, and their moles have the right to stand and be counted. Fortunately, there's a convenient way to convert between the moles of gaseous particles and their *volume.* Unlike gram atomic/molecular/formula masses, this conversion factor is constant *no matter what kinds of molecules make up the gas.* Every gas has a volume of 22.4 liters per mole, regardless of the size of the gaseous molecules.

Before you start your hooray-chemistry-is-finally-getting-simple dance, however, understand that certain conditions apply to this conversion factor. For example, it's only true at *standard temperature and pressure (STP),* or 0° Celsius and 1 atm. Also, the figure of 22.4L mol^{-1} applies only to the extent that a gas resembles an ideal gas, one whose particles have zero volume and neither attract nor repel one another. Ultimately, no gas is truly ideal, but many are so close to being so that the 22.4L mol^{-1} conversion is very useful.

What if you want to convert between the volume of a gaseous substance and its mass, or the mass of a substance and the number of particles it contains? You already have all the information you need! To make these kinds of conversions, simply build a chain of conversion factors, converting units step by step from the ones you have (say, liters) to the ones you want (say, grams). You'll find that your chain of conversion factors always includes central links featuring units of moles. You can think of the mole as a family member who passes on what you've said, loudly barking into the ear of your nearly deaf grandmother, because you have laryngitis and can't speak any louder. Without such a central translator, your message would no doubt be misinterpreted. "Grandma, how was your day?" would be received as "Grandma, you want to eat clay?" So, unless you're bent on force-feeding clay to your grandmother, do not attempt to convert directly from volume to mass, from mass to particles, or any other such shortcut. Use your translator, the mole.

Q. On her last day of work before retirement, Dr. Dentura daydreams about her newly purchased condo in Florida. Distracted, Dr. Dentura forgets to turn off the dinitrogen monoxide (laughing gas) after administering it to a root canal patient. The gas flows, filling 10 percent of the room, until her dental hygienist notices the hissing and nonchalantly turns off the gas. Being denser than air, the laughing gas settles near the floor so the dentist doesn't notice. How many moles of dinitrogen monoxide escaped into the room? Assume that Dr. Dentura suffers from frequent hot flashes and therefore keeps her office at a chilly 0°C. Further assume that her exam room is a spacious 10-x-10-x-10-foot cube.

A. **126 mol.** The problem tells you that the exam room is at standard temperature, and because the good doctor is presumably operating somewhere on the surface of the earth, it's safe to assume that the local pressure is somewhere around 1 atm. Having assured yourself that the room is at STP, you can safely use the 22.4L mol^{-1} conversion factor. To use it effectively, you must convert the volume of the room into liters. You know that the room has a volume of 10 ft × 10 ft × 10 ft = 1,000 ft^3. You also know that the dinitrogen oxide gas fills 10 percent of the room, and so accounts for 100 ft^3. Convert from cubic feet to liters, referring if you like to Chapter 2 for the conversion factor:

$$\frac{100\ \text{ft}^3}{1} \times \frac{(12\ \text{in})^3}{(1\ \text{ft})^3} \times \frac{(2.54\ \text{cm})^3}{(1\ \text{in})^3} \times \frac{1\ \text{mL}}{1\ \text{cm}^3} \times \frac{1\ \text{L}}{1000\ \text{mL}} = 2.83 \times 10^3 \text{L}$$

Then convert this volume in liters to moles by using the STP conversion factor of 22.4L mol^{-1}:

$$\frac{2.83 \times 10^3 \text{L}}{1} \times \frac{1\ \text{mol}}{22.4\text{L}} = 126\ \text{mol}$$

Q. Supervillain Lex Luthor accidentally purchases a vessel containing 0.05 kg of krypton, mistaking it for kryptonite, from an online retailer. Kryptonite, of course, is a glowing, green, and entirely fictional solid capable of utterly destroying Luthor's arch nemesis, Superman. Krypton, by contrast, is a relatively innocuous noble gas. What's the volume of Lex Luthor's unhelpful vessel, assuming that it was meticulously filled to only atmospheric pressure and is shipped on ice?

A. **13.4L.** You're given a mass and are asked to convert it to a volume. Ice keeps the vessel at a temperature near 0°C, and you're assured that the internal pressure of the vessel is 1 atm. So, you can assume STP. First, convert from grams to moles by using the gram atomic mass (in this case, 83.8g). Then convert from moles to volume, as shown in the following equation. The answer is 13.4 liters. That's nearly 4 gallons of useless noble gas, an embarrassing error for a supervillain.

$$\frac{0.05 \text{ kg}}{1} \times \frac{1000g}{1 \text{ kg}} \times \frac{1 \text{ mol}}{83.8g} \times \frac{22.4L}{1 \text{ mol}} = 13.4L$$

4. How many moles make up 350g of table salt, NaCl?

Solve It

5. The average volume of a human breath is roughly 500 mL. How many moles do you inhale with each breath on a day when the temperature hovers near freezing? Assuming that human exhalations eventually intermix evenly throughout the approximately 3×10^{22}L of Earth's atmosphere, calculate how many molecules from Genghis Khan's last breath you take into your lungs each time you inhale.

Solve It

6. In pounds, what is the mass of 3 moles of platinum?

Solve It

7. If you fill a 2L soda bottle (at STP) with carbon dioxide, how many particles does the bottle contain?

Giving Credit Where It's Due: Percent Composition

Chemists often are concerned with precisely what percentage of a compound's mass consists of one particular element. Lying awake at night, uttering prayers to Avogadro, they fret over this quantity, called *percent composition*. Calculating percent composition is trickier than you may think. Consider the following problem, for example.

The human body is composed of 60 to 70 percent water, and water contains twice as many hydrogen atoms as oxygen atoms. If two-thirds of every water molecule is hydrogen and water makes up 60 percent of the body, it seems logical to conclude that hydrogen makes up 40 percent of the body. Yet hydrogen is only the third most abundant element in the body *by mass*. What gives?

Oxygen is 16 times more massive than hydrogen. So, equating *atoms* of hydrogen and *atoms* of oxygen is a bit like equating a toddler to a sumo wrestler. When the doors of the elevator won't close, the sumo wrestler is the first one you should kick out, weep though he may.

Within a compound, it's important to sort out the atomic toddlers from the atomic sumo wrestlers. To do so, follow three simple steps.

1. **Calculate the gram molecular mass or gram formula mass of the compound, as we explain in the previous section.**

   Percent compositions are completely irrelevant to gram atomic masses, because these apply only to pure monatomic substances; by definition, these substance have 100 percent composition of a given element.

2. **Multiply the atomic mass of each element present in the compound by the number of atoms of that element present in one molecule.**

3. **Divide each of the masses calculated in Step 2 by the total mass calculated in Step 1. Multiply each fractional quotient by 100%. Voilà! You have the *percent composition* by mass of each element in the compound.**

Q. Calculate the percent composition for each element present in sodium sulfate, Na_2SO_4.

A. **Na: 32.4%, S: 22.5%, O: 45.1%.** The gram formula mass of sodium sulfate is

$(2 \times 23.0g\ mol^{-1}) + (1 \times 32.0g\ mol^{-1}) + (4 \times 16.0g\ mol^{-1}) = 142g\ mol^{-1}$.

Of each mole of compound,

$2 \times 23.0g = 46.0g$ are sodium

$1 \times 32.0g = 32.0g$ are sulfur

$4 \times 16.0g = 64.0g$ are oxygen

Dividing each of these quantities by the molar mass of sodium sulfate (142g) yields the percent composition:

$$\frac{46g}{142g} \times 100 = 32.4\%\ \text{sodium}, \quad \frac{32g}{142g} \times 100 = 22.5\%\ \text{sulfur, and} \quad \frac{64g}{142g} \times 100 = 45.1\%\ \text{oxygen}$$

As a check, add the three percentages to ensure they equal 100%: 32.4% + 22.5% + 45.1% = 100%.

8. Calculate the percent composition of potassium chromate, K_2CrO_4.

Solve It

9. Calculate the percent composition of propane, C_3H_8.

Solve It

Moving from Percent Composition to Empirical Formulas

What if you don't know the formula of a compound? Chemists sometimes find themselves in this disconcerting scenario. Rather than curse Avogadro (or perhaps *after* doing so), they analyze samples of the frustrating unknown to identify the percent composition. From there, they calculate the ratios of different types of atoms in the compound. They express these ratios as an *empirical formula,* the lowest whole number ratio of elements in a compound.

To find an empirical formula given percent composition, use the following procedure:

1. Assume that you have 100g of the unknown compound.

   The beauty of this little trick is that you conveniently gift yourself with the same number of grams of each elemental component as its contribution to the percent composition. For example, if you assume that you have 100g of a compound composed of 60.3% magnesium and 39.7% oxygen, you know that you have 60.3g of magnesium and 39.7g of oxygen.

2. Convert the assumed masses from Step 1 into moles by using gram atomic masses.

3. Divide each of the element-by-element mole quantities from Step 2 by the lowest among them.

   This division yields the mole ratios of the elements of the compound.

4. If any of your mole ratios aren't whole numbers, multiply all numbers by the smallest possible factor that produces whole number mole ratios for all the elements.

   For example, if there is 1 nitrogen atom for every 0.5 oxygen atom in a compound, the empirical formula is not $N_1O_{0.5}$. Such a formula casually suggests that an oxygen atom has been split, something that would create a small-scale nuclear explosion. Though impressive sounding, this scenario is almost certainly false. Far more likely is that the atoms of nitrogen to oxygen are combining in a 1:0.5 *ratio,* but do so in groups of $2 \times (1:0.5) = 2:1$. The empirical formula is thus N_2O.

   Because the original percent composition data is typically experimental, expect to see a bit of error in the numbers. For example, 2.03 is probably within experimental error of 2.

5. Write the empirical formula by attaching these whole-number mole ratios as subscripts to the chemical symbol of each element. Order the elements according to the general rules for naming ionic and molecular compounds (described in Chapter 6).

Q. What is the empirical formula of a substance that is 40.0% carbon, 6.7% hydrogen, and 53.3% oxygen, by mass?

A. **CH$_2$O.** For the sake of simplicity, assume that you have a total of 100g of this mystery compound. Therefore, you have 40.0g of carbon, 6.7g of hydrogen, and 53.3g of oxygen. Convert each of these masses to moles by using the gram atomic masses of C, H, and O.

$$\frac{40.0\text{g C}}{1} \times \frac{1 \text{ mol C}}{12\text{g C}} = 3.33 \text{ mol C}$$

$$\frac{6.7\text{g H}}{1} \times \frac{1 \text{ mol H}}{1\text{g H}} = 6.7 \text{ mol H}$$

$$\frac{53.3\text{g O}}{1} \times \frac{1 \text{ mol O}}{16\text{g O}} = 3.33 \text{ mol O}$$

Notice that the carbon and oxygen mole numbers are the same, so you know the ratio of these two elements is 1:1 within the compound. Next, divide all the mole numbers by the smallest among them, which is 3.33. This division yields

$$\frac{3.33 \text{ mol C}}{3.33} = 1 \text{ mol C}, \quad \frac{6.70 \text{ mol H}}{3.33} = 2 \text{ mol H}, \text{ and } \frac{3.33 \text{ mol O}}{3.33} = 1 \text{ mol O}$$

The compound has the empirical formula CH$_2$O. The actual number of atoms within each particle of the compound is some multiple of the numbers expressed in this formula.

---

10. Calculate the empirical formula of a compound with a percent composition of 88.9% oxygen and 11.1% hydrogen.

Solve It

11. Calculate the empirical formula of a compound with a percent composition of 40.0% sulfur and 60.0% oxygen.

Solve It

Moving from Empirical Formulas to Molecular Formulas

Many compounds in nature, particularly compounds made of carbon, hydrogen, and oxygen, are composed of atoms that occur in numbers that are multiples of their empirical formula. In other words, their empirical formulas don't reflect the actual numbers of atoms within them, but only the ratios of those atoms. What a nuisance! Fortunately, this is an old nuisance, so chemists have devised a means to deal with it. To account for these annoying types of compounds, chemists are careful to differentiate between an empirical formula and a *molecular formula*. A molecular formula uses subscripts that report the actual number of atoms of each type in a molecule of the compound (a *formula unit* accomplishes the same thing for ionic compounds).

Molecular formulas are associated with gram molecular masses that are simple whole number multiples of the corresponding *empirical formula mass*. In other words, a molecule with the empirical formula CH_2O has an empirical formula mass of $30.0g\ mol^{-1}$ (12 for the carbon + 2 for the two hydrogens + 16 for the oxygen). The molecule may have a molecular formula of CH_2O, $C_2H_4O_2$, $C_3H_6O_3$, and so on. As a result, the compound may have a gram molecular mass of $30.0g\ mol^{-1}$, $60.0g\ mol^{-1}$, $90.0g\ mol^{-1}$, and so on.

You can't calculate a molecular formula based on percent composition alone. If you attempt to do so, Avogadro and Perrin will rise from their graves, find you, and slap you 6.022×10^{23} times per cheek (Ooh, that smarts!). The folly of such an approach is made clear by comparing formaldehyde with glucose. The two compounds have the same empirical formula, CH_2O, but different molecular formulas, CH_2O and $C_6H_{12}O_6$, respectively. Glucose is a simple sugar, the one made by photosynthesis and the one broken down during cellular respiration. You can dissolve it into your coffee with pleasant results. Formaldehyde is a carcinogenic component of smog. Solutions of formaldehyde have historically been used to embalm dead bodies. You are not advised to dissolve formaldehyde into your coffee. In other words, molecular formulas differ from empirical formulas, and the difference is important.

To determine a molecular formula, you must know the gram formula mass of the compound as well as the empirical formula (or enough information to calculate it yourself from the percent composition; see the previous section for details). With these tools in hand, calculating the molecular formula involves a three-step process:

1. **Calculate the empirical formula mass.**

2. **Divide the gram molecular mass by the empirical formula mass.**

3. **Multiply each of the subscripts within the empirical formula by the number calculated in Step 2.**

Q. What is the molecular formula of a compound which has a gram molecular mass of 34g mol^{-1} and the empirical formula HO.

A. H$_2$O$_2$. The empirical formula mass is

$$1g\ mol^{-1} + 16g\ mol^{-1} = 17g\ mol^{-1}.$$

Dividing the gram molecular mass by this value yields

$$34g\ mol^{-1}/17g\ mol^{-1} = 2.$$

Multiplying the subscripts within the empirical formula by this number gives the molecular formula H$_2$O$_2$. This formula corresponds to the compound hydrogen peroxide.

12. What is the molecular formula of a compound which has a gram formula mass of 78g mol^{-1} and the empirical formula NaO.

Solve It

13. A compound has a percent composition of 49.5% carbon, 5.2% hydrogen, 16.5% oxygen, and 28.8% nitrogen. The compound's gram molecular mass is 194.2g mol^{-1}. What are the empirical and molecular formulas?

Solve It

Answers to Questions on Moles

The following are the answers to the practice problems presented in this chapter.

1 **0.03 mol.** Here's how you calculate the answer (remember Chapter 1's rules for multiplying and dividing in scientific notation):

$$\frac{1.5\times10^{11}\text{ stars}}{1\text{ galaxy}}\times\frac{1.25\times10^{11}\text{galaxies}}{\text{universe}}\times\frac{1\text{ mol stars}}{6.022\times10^{23}\text{stars}}=0.03\text{ mol}$$

Reflect on this answer. The entire universe contains only 3 percent of a mole of stars!

2 **2×10^{-6} mol.** Here's the calculation:

$$\frac{7\times10^{9}\text{ people}}{1}\times\frac{2\times10^{8}\text{ insects}}{1\text{ person}}\times\frac{1\text{ mol}}{6.022\times10^{23}\text{ insects}}=0.000002\text{ mol}$$

Again, a number that nonchemists find unimaginably large comes out to the tiniest fraction of 1 mole.

3 **1.72×10^{15} atoms.** First, find the total mass of carbon in the body by taking 23% of 150 kg, which equals 34.5 kg. Then convert those 34.5 kg to atoms by using the conversion factors given in the problem and the conversion between moles and atoms (which you already know):

$$\frac{34.5\text{ kg}}{1}\times\frac{83\text{ mol}}{1\text{ kg}}\times\frac{6.022\times10^{23}\text{ atoms C}}{1\text{ mol C}}\times\frac{1\text{ atom 14C}}{1\times10^{12}\text{ atoms C}}=1.72\times10^{15}\text{ atoms}$$

4 **6.0 mol.** First, find the gram formula mass of sodium chloride using the same steps you used to calculate gram molecular mass. It's 23g mol^{-1} + 35.5g mol^{-1} = 58.5g mol^{-1}. Given the gram formula mass, the conversion is simple:

$$\frac{350\text{g NaCl}}{1}\times\frac{1\text{ mol NaCl}}{58.5\text{g NaCl}}=6.0\text{ mol}$$

5 **0.02 mol; 0.2 particles.** First, calculate how many moles of air you inhale with each breath:

$$\frac{0.5\text{L}}{1}\times\frac{1\text{ mol}}{22.4\text{L}}=0.02\text{ mol}$$

Next, calculate the number of particles in each mole of air that were once part of Genghis Khan's final breath. Do this by finding the fraction of the atmosphere comprised by a single breath. Then, multiply that number by the number of particles in 1 mole:

$$\frac{0.5\text{L}}{3\times10^{22}\text{L}}\times\frac{6.022\times10^{23}\text{particles}}{1\text{ mol}}=10\text{ particles per mole}$$

That's right — 10 particles in every mole of air on the planet were inhaled by the mighty Mongolian in his last scowling moment. Now, figure out how many of these particles you take in when you inhale. Multiply the total number of inhaled moles by the number of Genghis-kissed particles per mole:

$$\frac{0.02 \text{ mol}}{1} \times \frac{10 \text{ particles}}{1 \text{ mol}} = 0.2 \text{ particles}$$

This means that, on average, one out of every five times you inhale, you breathe in a piece of East Asian history!

6 **1.3 lb.** Use your conversion factor techniques from Chapter 1 to convert moles to grams to pounds as shown in the following equation:

$$\frac{3 \text{ mol Pt}}{1} \times \frac{195 \text{g Pt}}{1 \text{ mol Pt}} \times \frac{1 \text{ lb}}{453 \text{g}} = 1.3 \text{ lb}$$

7 $\mathbf{5.37 \times 10^{22}}$ **particles.** Use conversion factors to convert from liters to moles, and then moles to particles as shown in the following equation:

$$\frac{2 \text{L}}{1} \times \frac{1 \text{ mol}}{22.4 \text{L}} \times \frac{6.022 \times 10^{23} \text{particles}}{1 \text{ mol}} = 5.37 \times 10^{22} \text{particles}$$

8 **K: 40.3%; Cr: 26.8%; O: 33.0%.** First, calculate the gram molecular mass of potassium chromate, which comes to 194.2g mol^{-1}:

$$(2 \times 39.1 \text{g mol}^{-1}) + (1 \times 52.0 \text{g mol}^{-1}) + (4 \times 16.0 \text{g mol}^{-1}) = 194.2 \text{g mol}^{-1}$$

In a 194.2g sample, 78.2g are potassium, 52.0g are chromium, and 64.0g are oxygen. Divide each of these masses by the gram molecular mass, and multiply by 100 to get the percent composition. *Note:* If you've rounded your percentages properly, they don't quite add to 100%, but rather 100.1%. If you had done away with rounding, you would have gotten exactly 100%. Rounding is common practice though, so don't be too worried if your answer is off by a tenth or two.

$$\frac{78.2 \text{g}}{194.2 \text{g}} \times 100 = 40.3\% \text{ potassium}, \quad \frac{52.0 \text{g}}{194.2 \text{g}} \times 100 = 26.8\% \text{ chromium}, \text{ and } \frac{64.0 \text{g}}{194.2 \text{g}} \times 100 = 33.0\% \text{ oxygen}$$

9 **C: 81.8%; H: 18.2%.** The molecular mass of propane is

$$(3 \times 12.0 \text{g mol}^{-1}) + (8 \times 1.0 \text{g mol}^{-1}) = 44 \text{g mol}^{-1}$$

The percent composition of propane is therefore:

$$\frac{36.0 \text{g}}{44.0 \text{g}} \times 100 = 81.8\% \text{ carbon and } \frac{8.0 \text{g}}{44.0 \text{g}} \times 100 = 18.2\% \text{ hydrogen}$$

10 $\mathbf{H_2O}$**.** First, assume that you have 88.9g of oxygen and 11.1g of hydrogen in a 100g sample. Then convert each of these masses into moles by using the gram atomic masses of oxygen and hydrogen:

$$\frac{88.9 \text{g O}}{1} \times \frac{1 \text{ mol O}}{16.0 \text{g O}} = 5.55 \text{ mol}$$

$$\frac{11.1 \text{g H}}{1} \times \frac{1 \text{ mol H}}{1.0 \text{g H}} = 11.1 \text{ mol}$$

Next, divide each of these mole quantities by the smallest among them, 5.55 mol:

$$\frac{5.55 \text{ mol O}}{5.55} = 1 \text{ mol O and } \frac{11.1 \text{ mol H}}{5.55} = 2 \text{ mol H}$$

Attach these quotients as subscripts and list the atoms properly. This yields H_2O. The compound is water.

11 **SO_3.** Following the same procedure as in Question 10, you calculate 1.25 mol sulfur and 3.75 mol oxygen. Dividing each of these quantities by 1.25 mol (the smallest quantity) yields 1.25 / 1.25 = 1 mol of sulfur and 3.75 / 1.25 = 3 mol oxygen, or a mole ratio of 1:3. The compound is SO_3, sulfur trioxide.

12 **Na_2O_2.** First, find the empirical formula mass of NaO, which is

$(1 \times 23.0 \text{g mol}^{-1}) + (1 \times 16.0 \text{g mol}^{-1}) = 39.0 \text{g mol}^{-1}$

Then divide the gram formula mass of the mystery compound, 78g mol^{-1}, by this empirical formula mass to obtain the quotient, 2. Multiply each of the subscripts within the empirical formula by this number to obtain Na_2O_2. You've just found the molecular formula for sodium peroxide.

13 **$C_4H_5N_2O$** is the empirical formula; **$C_8H_{10}N_4O_2$** is the molecular formula. You're not directly given the empirical formula of this compound. But you *are* given the percent composition. Using the percent composition, you can calculate the empirical formula. To do this, assume that you have 100g of the substance, giving you 49.5g carbon, 5.2g hydrogen, 16.5g oxygen, and 28.8g nitrogen. Then divide these masses by the atomic mass of each element, giving you

49.5 / 12.0 = 4.125 mol carbon

5.2 / 1.0g = 5.2 mol hydrogen

16.5 / 16.0 = 1.031 mol oxygen

28.8 / 14.0 = 2.057 mol nitrogen

Finally, divide each of these mole values by the lowest among them, 1.031, giving you 4.0 mol carbon, 5.0 mol hydrogen, 1.0 mol oxygen, and 2.0 mol nitrogen, giving you the empirical formula $C_4H_5N_2O$.

Here's how you get the molecular formula: The empirical formula mass is 97.1g mol^{-1} (calculated by multiplying the number of atoms of each element in the compound by its atomic mass and adding them all up). Dividing the gram molecular mass you were given (194.2g mol^{-1}) by this empirical formula mass yields the quotient, 2. Multiplying each of the subscripts in the empirical formula by 2 produces the molecular formula, $C_8H_{10}N_4O_2$. The common name for this culturally important compound is caffeine.

Chapter 8

Getting a Grip on Chemical Equations

● ●

In This Chapter

▶ Reading, writing, and balancing chemical equations

▶ Recognizing five types of reactions and predicting products

▶ Charging through net ionic equations

● ●

C hapters 5, 6, and 7 focus on chemical compounds and the bonds that bind them. You can think of a compound in two different ways:

✔ As the product of one chemical reaction

✔ As a starting material in another chemical reaction

In the end, chemistry is about action — about the breaking and making of bonds. Chemists describe action by using *chemical equations,* sentences that say who reacted with whom, and who remained when the smoke cleared. This chapter explains how to read, write, balance, and predict the products of these action-packed chemical sentences.

Translating Chemistry into Equations and Symbols

In general, all chemical equations are written in the basic form

Reactants → Products

where the arrow in the middle means *yields.* The basic idea is that the reactants react, and the reaction produces products. By *reacting,* we simply mean that bonds within the reactants are broken, to be replaced by new and different bonds within the products.

Chemists fill chemical equations with symbols because they think it looks cool and, more importantly, because the symbols help pack a lot of meaning into a small space. Table 8-1 summarizes the most important symbols you'll find in chemical equations.

Table 8-1	Symbols Commonly Used in Chemical Equations
Symbol	**Explanation**
+	Separates two reactants or products.
→	The "yields" symbol separates the reactants from the products. The single arrowhead suggests the reaction occurs in only one direction.
↔	A two-way yields symbol means the reaction can occur reversibly, in both directions. You may also see this symbol written as two stacked arrows with opposing arrowheads.
(s)	A reactant or product followed by this symbol exists as a solid.
(l)	A reactant or product followed by this symbol exists as a liquid.
(g)	A reactant or product followed by this symbol exists as a gas.
(aq)	A reactant or product followed by this symbol exists in aqueous solution, dissolved in water.
Δ	This symbol, usually written above the yields symbol, signifies that heat is added to the reactants.
Ni, LiCl	Sometimes a chemical symbol (such as those for nickel or lithium chloride here) is written above the yields symbol. This means that the indicated chemical was added as a catalyst. Catalysts speed up reactions but do not otherwise participate in them.

After you understand how to interpret chemical symbols, compound names (see Chapter 6), and the symbols in Table 8-1, there's not a lot you can't understand. You're equipped, for example, to decode a chemical equation into an English sentence describing a reaction. Conversely, you can translate an English sentence into the chemical equation it describes. When you're fluent in this language, you regrettably won't be able to talk to the animals; you will, however, be able to describe their metabolism in great detail.

Q. Write out the chemical equation for the following sentence:

Solid iron (III) oxide reacts with gaseous carbon monoxide to produce solid iron and gaseous carbon dioxide.

A. $Fe_3O_2(s) + CO(g) \rightarrow Fe(s) + CO_2(g)$. First, convert each formula into the written name for the compound. Next, annotate the physical state of the compound, if it's provided. Next, group the compounds into "reactant" and "product" categories. Things that react are reactants. Things that are produced are products. List the reactants on the left side of a reaction arrow, separating each with a plus sign. Do the same for the products, but list them on the right side of the reaction arrow.

Q. Write a sentence that describes the following chemical reaction:

$$H_2O(l) + N_2O_3(g) \xrightarrow{\Delta} HNO_2(aq)$$

A. **Liquid water is heated with gaseous dinitrogen trioxide to produce an aqueous solution of nitrous acid.** First, figure out the names of the compounds. Next, note their states (liquid, solid, gas, aqueous solution). Then observe what the compounds are actually doing — combining, decomposing, combusting, and so on. Finally, assemble all these observations into a sentence. Many sentences could be correct, as long as they include these elements.

1. Write chemical equations for the following reactions:

  a. Solid magnesium is heated with gaseous oxygen to form solid magnesium oxide.

  b. Solid diboron trioxide reacts with solid magnesium to make solid boron and solid magnesium oxide.

Solve It

2. Write sentences describing the reactions summarized by the following chemical equations:

  a. $S(s) + O_2(g) \rightarrow SO_2(g)$

  b. $H_2(g) + O_2(g) \xrightarrow{Pt} H_2O(l)$

Solve It

Making Your Chemical Equations True by Balancing

The equations you read and wrote in the previous section are *skeleton equations,* and are perfectly adequate for a qualitative description of the reaction: who are the reactants and who are the products. But if you look closely, you'll see that those equations just don't add up quantitatively. As written, the mass of one mole of each of the reactants doesn't equal the mass of one mole of each of the products (see Chapter 7 for details on moles). The skeleton equations break the *Law of Conservation of Mass,* which states that all the mass present at the beginning of a reaction must be present at the end. To be quantitatively accurate, these equations must be *balanced* so the masses of reactants and products are equal.

To balance an equation, you use *coefficients* to alter the number of moles of reactants and/or products so the mass on one side of the equation equals the mass on the other side. A *coefficient* is simply a number that precedes the symbol of an element or compound, multiplying the number of moles of that *entire* compound within the equation. Coefficients are different from *subscripts,* which multiply the number of atoms or groups within a compound. Consider the following:

    $4Cu(NO_3)_2$

The number 4 that precedes the compound is a coefficient, indicating that there are four moles of copper (II) nitrate. The subscripted 3 and 2 within the compound indicate that each nitrate contains three oxygen atoms, and that there are two nitrate groups per atom of copper. Coefficients and subscripts multiply to yield the total number of moles of each atom:

$$4 \text{ mol } Cu(NO_3)_2 \times (1 \text{ mol } Cu/\text{mol } Cu(NO_3)_2) = \textbf{4 mol Cu}$$

$$4 \text{ mol } Cu(NO_3)_2 \times (2 \text{ mol } NO_3/\text{mol } Cu(NO_3)_2) \times 1 \text{ mol } N/\text{mol } NO_3 = \textbf{8 mol N}$$

$$4 \text{ mol } Cu(NO_3)_2 \times (2 \text{ mol } NO_3/\text{mol } Cu(NO_3)_2) \times 3 \text{ mol } O/\text{mol } NO_3 = \textbf{24 mol O}$$

When you balance an equation, *you change only the coefficients.* Changing subscripts alters the chemical compounds themselves. If your pencil were equipped with an electrical shocking device, that device would activate the moment you attempted to change a subscript while balancing an equation.

TIP

Here's a simple recipe for balancing equations:

1. **Given a skeleton equation (one that includes formulas for reactants and products), count up the number of each kind of atom on each side of the equation.**

   If you recognize any polyatomic ions, you can count these as one whole group (as if they were their own form of element). See Chapter 6 for information on recognizing polyatomic ions.

2. **Use coefficients to balance the elements or polyatomic ions, one at a time.**

   For simplicity, start with those elements or ions that appear only once on each side.

3. **Check the equation to ensure that each element or ion is balanced.**

   Checking is important because you may have "ping-ponged" several times from reactants to products and back — there's plenty of opportunity for error.

4. **When you're sure the reaction is balanced, check to make sure it's in lowest terms.**

   For example,

   $$4H_2(g) + 2O_2(g) \rightarrow 4H_2O(l)$$

   should be reduced to

   $$2H_2(g) + O_2(g) \rightarrow 2H_2O(l)$$

EXAMPLE

Q. Balance the following equation:

$$Na(s) + Cl_2(g) \rightarrow NaCl(s)$$

A. $2Na(s) + Cl_2(g) \rightarrow 2NaCl(s)$. You can't change the subscripted 2 in the chlorine gas reactant, so you must add a coefficient of 2 to the sodium chloride product. This change requires you to balance the sodium reactant with another coefficient of 2.

3. Balance the following reactions:

a. $N_2(g) + H_2(g) \rightarrow NH_3(g)$

b. $C_3H_8(g) + O_2(g) \rightarrow CO_2(g) + H_2O(l)$

c. $Al(s) + O_2(g) \rightarrow Al_2O_3(s)$

d. $AgNO_3(aq) + Cu(s) \rightarrow Cu(NO_3)_2(aq) + Ag(s)$

Solve It

4. Balance the following reactions:

a. $Ag_2SO_4(aq) + AlCl_3(aq) \rightarrow AgCl(s) + Al_2(SO_4)_3(aq)$

b. $CH_4(g) + Cl_2(g) \rightarrow CH_2Cl_2(l) + HCl(g)$

c. $Cu(s) + HNO_3(aq) \rightarrow Cu(NO_3)_2(aq) + NO(g) + H_2O(l)$

d. $HCl(aq) + Ca(OH)_2(aq) \rightarrow CaCl_2(s) + H_2O(l)$

Solve It

5. Balance the following reactions:

a. $H_2C_2O_4(aq) + KOH(aq) \rightarrow K_2C_2O_4(aq) + H_2O(l)$

b. $P(s) + Br_2(l) \rightarrow PBr_3(g)$

c. $Pb(NO_3)_2(aq) + KI(aq) \rightarrow KNO_3(aq) + PbI_2(s)$

d. $Zn(s) + AgNO_3(aq) \rightarrow Zn(NO_3)_2(aq) + Ag(s)$

Solve It

6. Balance the following reactions:

a. $NaOH(aq) + H_2SO_4(aq) \rightarrow H_2O(l) + Na_2SO_4(aq)$

b. $HNO_3(aq) + S(s) \rightarrow SO_2(g) + NO(g) + H_2O(l)$

c. $CuO(s) + H_2(g) \rightarrow Cu(s) + H_2O(l)$

d. $KMnO_4(aq) + HCl(aq) \rightarrow MnCl_2(aq) + Cl_2(g) + H_2O(l) + KCl(aq)$

Solve It

Recognizing Reactions and Predicting Products

You can't begin to wrap your brain around the unimaginably large number of possible chemical reactions. That's good that so many reactions can occur, because they make things like life and the universe possible. From the perspective of a mere human brain trying to grok all these reactions, we have another bit of good news: A few categories of reactions pop up over and over again. After you see the very basic patterns in these categories, you'll be able to make sense of the majority of reactions out there.

The following sections describe five types of reactions that you'd do well to recognize (notice how their names tell you what happens in each reaction). By recognizing the patterns of these five types of reactions, you can often predict reaction products given only a set of reactants. There are no perfect guidelines, and predicting reaction products can take what is called *chemical intuition,* a sense of what reaction is likely to occur based on knowing the outcomes of similar reactions. Still, if you're given both reactants and products, you should be able to tell what kind of reaction connects them, and if you're given reactants and the type of reaction, you should be able to predict likely products. Figuring out the formulas of products often requires you to apply knowledge about how ionic and molecular compounds are put together. To review these concepts, see Chapters 5 and 6.

Combination

Two or more reactants combine to form a single product, following the general pattern

$$A + B \rightarrow C$$

For example,

$$2Na(s) + Cl_2(g) \rightarrow 2NaCl(s)$$

The combining of elements to form compounds (like NaCl) is a particularly common kind of combination reaction. Here is another example:

$$2Ca(s) + O_2(g) \rightarrow 2CaO(s)$$

Compounds can also combine to form new compounds, such as in the combination of sodium oxide with water to form sodium hydroxide:

$$Na_2O(s) + H_2O(l) \rightarrow 2NaOH(aq)$$

Decomposition

A single reactant breaks down (decomposes) into two or more products, following the general pattern

$$A \rightarrow B + C$$

For example,

$$2H_2O(l) \rightarrow 2H_2(g) + O_2(g)$$

Notice that combination and decomposition reactions are the same reaction in opposite directions.

Many decomposition reactions produce gaseous products, such as in the decomposition of carbonic acid into water and carbon dioxide:

$$H_2CO_3(aq) \rightarrow H_2O(l) + CO_2(g)$$

Single replacement

In a single replacement reaction, a single, more reactive element or group replaces a less reactive element or group, following the general pattern

$$A + BC \rightarrow AC + B$$

For example,

$$Zn(s) + CuSO_4(aq) \rightarrow ZnSO_4(aq) + Cu(s)$$

Single replacement reactions in which metals replace other metals are especially common. You can determine which metals are likely to replace which others by using the *metal activity series,* a ranked list of metals in which ones higher on the list tend to replace ones lower on the list. Table 8-2 presents the metal activity series.

Table 8-2	Metal Activity Series
Metal	*Notes*
Lithium Potassium Strontium Calcium Sodium	Most reactive metals; react with cold water to form hydroxide and hydrogen gas.
Magnesium Aluminum Zinc Chromium	React with hot water/acid to form oxides and hydrogen gas.
Iron Cadmium Cobalt Nickel Tin Lead	Replace hydrogen ion from dilute strong acids.
Hydrogen	Non-metal, listed in reactive order.
Antimony Arsenic Bismuth Copper	Combine directly with oxygen to form oxides.

(continued)

Table 8-2 *(continued)*	
Metal	**Notes**
Mercury Silver Palladium Platinum Gold	Least reactive metals; often found as free metals; oxides decompose easily.

Double replacement

Double replacement is a special form of *metathesis reaction* (that is, a reaction in which two reacting species exchange bonds). Double replacement reactions tend to occur between ionic compounds in solution. In these reactions, cations (atoms or groups with positive charge) from each reactant swap places to form ionic compounds with the opposing anions (atoms or groups with negative charge), following the general pattern

$$AB + CD \rightarrow AD + CB$$

For example,

$$KCl(aq) + AgNO_3(aq) \rightarrow AgCl(s) + KNO_3(aq)$$

Of course, ions dissolved in solution move about freely, not as part of cation-anion complexes. So, to allow double replacement reactions to progress, one of several things must occur.

✔ One of the product compounds must be insoluble, so it *precipitates* (forms an insoluble solid) out of solution after it forms.

✔ One of the products must be a gas that bubbles out of solution after it forms.

✔ One of the products must be a solvent molecule, such as H_2O, that separates from the ionic compounds after it forms.

Combustion

Oxygen is always a reactant in combustion reactions, which often release heat and light as they occur. Combustion reactions frequently involve hydrocarbon reactants (like propane, $C_3H_8(g)$, the gas used to fire up backyard grills), and yield carbon dioxide and water as products. For example,

$$C_3H_8(g) + 5O_2(g) \rightarrow 3CO_2(g) + 4H_2O(l)$$

Combustion reactions also include combination reactions between elements and oxygen, such as:

$$S(s) + O_2(g) \rightarrow SO_2(g)$$

So, if the reactants include oxygen (O_2) and a hydrocarbon or an element, you're probably dealing with a combustion reaction. If the products are carbon dioxide and water, you're almost certainly dealing with a combustion reaction.

Q. Complete and balance the following reaction:

$$Be(s) + O_2(g) \rightarrow$$

A. **2Be(s) + O₂(g) → 2BeO(s).** Although beryllium isn't on the metal activity series, you can make a pretty good prediction that this is a combination/combustion reaction. Why? First, you have two reactants. A single reactant would imply decomposition. The beryllium reactant is an element, not a compound, so you can rule out double replacement. The metal element reactant might make you consider single replacement. But there's no metal in dioxygen, the other reactant, so there's no obvious replacement partner. So the most likely reaction is combination. When elements combine with oxygen, that's also a combustion reaction.

---

7. Complete and balance the following reactions:

a. $C(s) + O_2(g) \rightarrow$

b. $Mg(s) + I_2(g) \rightarrow$

c. $Sr(s) + I_2(g) \rightarrow$

d. $Mg(s) + N_2(g) \rightarrow$

Solve It

8. Complete and balance the following reactions:

a. $HI(g) \rightarrow$

b. $H_2O_2(l) \rightarrow H_2O(l) +$

c. $NaCl(s) \rightarrow$

d. $MgCl_2(s) \rightarrow$

Solve It

9. Complete and balance the following reactions:

a. $Zn(s) + H_2SO_4(aq) \rightarrow$

b. $Al(s) + HCl(aq) \rightarrow$

c. $Li(s) + H_2O(l) \rightarrow$

d. $Mg(s) + Zn(NO_3)_2(aq) \rightarrow$

Solve It

10. Complete and balance the following reactions:

a. $Ca(OH)_2(aq) + HCl(aq) \rightarrow$

b. $HNO_3(aq) + NaOH(aq) \rightarrow$

c. $FeS(s) + H_2SO_4(aq) \rightarrow$

d. $BaCl_2(aq) + K_2CO_3(aq) \rightarrow$

Solve It

Getting Rid of Mere Spectators: Net Ionic Equations

Chemistry is often conducted in aqueous solutions. Soluble ionic compounds dissolve into their component ions, and these ions can react to form new products. In these kinds of reactions, sometimes only the cation or anion of a dissolved compound reacts. The other ion merely watches the whole affair, twiddling its charged thumbs in electrostatic boredom. These uninvolved ions are called *spectator ions.*

Because spectator ions don't actually participate in the chemistry of a reaction, you don't need to include them in a chemical equation. Doing so leads to a needlessly complicated reaction equation. So, chemists prefer to write *net ionic equations,* which omit the spectator ions. A net ionic equation doesn't include every component that may be present in a given beaker. Rather, it includes only those components that actually react.

Here is a simple recipe for making net ionic equations of your own:

1. **Examine the starting equation to determine which ionic compounds are dissolved, as indicated by the (*aq*) symbol following the compound name.**

   $Zn(s) + HCl(aq) \rightarrow ZnCl_2(aq) + H_2(g)$

2. **Rewrite the equation, explicitly separating dissolved ionic compounds into their component ions.**

   This step requires you to recognize common polyatomic ions, so be sure to familiarize yourself with them (flip to Chapter 6 for details).

   $Zn(s) + H^+(aq) + Cl^-(aq) \rightarrow Zn^{2+}(aq) + 2Cl^-(aq) + H_2(g)$

3. **Compare the reactant and product sides of the rewritten reaction. Any dissolved ions that appear in the same form on both sides are spectator ions. Cross out the spectator ions to produce a net reaction.**

   $Zn(s) + H^+(aq) + \cancel{Cl^-(aq)} \rightarrow Zn^{2+}(aq) + \cancel{2Cl^-(aq)} + H_2(g)$

   Net reaction:

   $Zn(s) + H^+(aq) \rightarrow Zn^{2+}(aq) + H_2(g)$

   As written, the preceding reaction is imbalanced with respect to the number of hydrogen atoms and the amount of positive charge.

4. **Balance the net reaction for mass and charge.**

   $Zn(s) + 2H^+(aq) \rightarrow Zn^{2+}(aq) + H_2(g)$

If you want, you can balance the equation for mass and charge first (at Step 1). This way, when you cross out spectator ions at Step 3, you cross out equivalent numbers of ions. Either method produces the same net ionic equation in the end. Some people prefer to balance the starting reaction equation, but others prefer to balance the net reaction because it's a simpler equation.

Q. Generate a balanced net ionic equation for the following reaction:

$CaCO_3(s) + 2HCl(aq) \rightarrow CaCl_2(aq) + H_2O(l) + CO_2(g)$

A. $CaCO_3(s) + 2H^+(aq) \rightarrow Ca^{2+}(aq) + H_2O(l) + CO_2(g)$

Because HCl and $CaCl_2$ are listed as aqueous (aq), rewrite the equation, explicitly separating those compounds into their ionic components:

$CaCO_3(s) + 2H^+(aq) + 2Cl^-(aq) \rightarrow Ca^{2+}(aq) + 2Cl^-(aq) + H_2O(l) + CO_2(g)$

Next, cross out any components that appear in the same form on both sides of the equation. In this case, the chloride ions (Cl^-) are crossed out:

$CaCO_3(s) + 2H^+(aq) + \cancel{2Cl^-(aq)} \rightarrow Ca^{2+}(aq) + \cancel{2Cl^-(aq)} + H_2O(l) + CO_2(g)$

This leaves the net reaction:

$CaCO_3(s) + 2H^+(aq) \rightarrow Ca^{2+}(aq) + H_2O(l) + CO_2(g)$

The net reaction turns out to be balanced for mass and charge, so it is the balanced net ionic equation.

---

11. Generate balanced net ionic equations for the following reactions:

a. $LiOH(aq) + HI(aq) \rightarrow H_2O(l) + LiI(aq)$

b. $AgNO_3(aq) + NaCl(aq) \rightarrow AgCl(s) + NaNO_3(aq)$

c. $Pb(NO_3)_2(aq) + H_2SO_4(aq) \rightarrow PbSO_4(s) + HNO_3(aq)$

Solve It

12. Generate balanced net ionic equations for the following reactions:

a. $HCl(aq) + ZnS(aq) \rightarrow H_2S(g) + ZnCl_2(aq)$

b. $Ca(OH)_2(aq) + H_3PO_4(aq) \rightarrow Ca_3(PO_4)_2(aq) + H_2O(l)$

c. $(NH_4)_2S(aq) + Co(NO_3)_2(aq) \rightarrow CoS(s) + NH_4NO_3(aq)$

Solve It

Answers to Questions on Chemical Equations

Chemistry is about action, the breaking and making of bonds. You've read the chapter, and you've answered the questions. Now check your answers to see whether any of the chemistry concepts in this chapter acted on your brain.

1 Don't forget the symbols indicating state, and whether any heat or catalyst were added.

a. $Mg(s) + O_2(g) \xrightarrow{\Delta} MgO(s)$

b. $B_2O_3(s) + Mg(s) \rightarrow B(s) + MgO(s)$

2 Here are the sentences describing the provided reactions:

a. Solid sulfur and gaseous oxygen react to produce the gas sulfur dioxide.

b. Hydrogen gas reacts with oxygen gas in the presence of platinum to produce liquid water.

3 Here are the balanced reactions:

a. $N_2(g) + 3H_2(g) \rightarrow 2NH_3(g)$

b. $C_3H_8(g) + 5O_2(g) \rightarrow 3CO_2(g) + 4H_2O(l)$

c. $4Al(s) + 3O_2(g) \rightarrow 2Al_2O_3(s)$

d. $2AgNO_3(aq) + Cu(s) \rightarrow Cu(NO_3)_2(aq) + 2Ag(s)$

4 More balanced reactions:

a. $3Ag_2SO_4(aq) + 2AlCl_3(aq) \rightarrow 6AgCl(s) + Al_2(SO_4)_3(aq)$

b. $CH_4(g) + 2Cl_2(g) \rightarrow CH_2Cl_2(l) + 2HCl(g)$

c. $3Cu(s) + 8HNO_3(aq) \rightarrow 3Cu(NO_3)_2(aq) + 2NO(g) + 4H_2O(l)$

d. $2HCl(aq) + Ca(OH)_2(aq) \rightarrow CaCl_2(s) + 2H_2O(l)$

5 Still more balanced reactions:

a. $H_2C_2O_4(aq) + 2KOH(aq) \rightarrow K_2C_2O_4(aq) + 2H_2O(l)$

b. $2P(s) + 3Br_2(l) \rightarrow 2PBr_3(g)$

c. $Pb(NO_3)_2(aq) + 2KI(aq) \rightarrow 2KNO_3(aq) + PbI_2(s)$

d. $Zn(s) + 2AgNO_3(aq) \rightarrow Zn(NO_3)_2(aq) + 2Ag(s)$

6 A final batch of balanced reactions:

a. $2NaOH(aq) + H_2SO_4(aq) \rightarrow 2H_2O(l) + Na_2SO_4(aq)$

b. $4HNO_3(aq) + 3S(s) \rightarrow 3SO_2(g) + 4NO(g) + 2H_2O(l)$

c. $CuO(s) + H_2(g) \rightarrow Cu(s) + H_2O(l)$

d. $2KMnO_4(aq) + 16HCl(aq) \rightarrow 2MnCl_2(aq) + 5Cl_2(g) + 8H_2O(l) + 2KCl(aq)$

7 After completing the reaction, it can be easy to forget to balance it, too. Of course, proper balancing means that you have to pay attention to the amount of each atom in the product compounds. These were all combination reactions:

    a. $C(s) + O_2(g) \rightarrow CO_2(g)$

    b. $Mg(s) + I_2(g) \rightarrow MgI2(s)$

    c. $Sr(s) + I_2(g) \rightarrow SrI_2(s)$

    d. $3Mg(s) + N_2(g) \rightarrow Mg_3N_2(s)$

8 Here are some more completed, balanced reactions. All of these were decomposition reactions:

    a. $2HI(g) \rightarrow H_2(g) + I_2(g)$

    b. $2H_2O_2(l) \rightarrow 2H_2O(l) + O_2(g)$

    c. $2NaCl(s) \rightarrow 2Na(s) + Cl_2(g)$

    d. $MgCl_2(s) \rightarrow Mg(s) + Cl_2(g)$

9 Here are the completed, balanced versions of a series of single replacement reactions:

    a. $Zn(s) + H_2SO_4(aq) \rightarrow ZnSO_4(aq) + H_2(g)$

    b. $2Al(s) + 6HCl(aq) \rightarrow 2AlCl_3(aq) + 3H_2(g)$

    c. $2Li(s) + 2H_2O(l) \rightarrow 2LiOH(aq) + H_2(g)$

    d. $Mg(s) + Zn(NO_3)_2(aq) \rightarrow Mg(NO_3)_2(aq) + Zn(s)$

10 This final set, completed and balanced, is composed of double replacement reactions:

    a. $Ca(OH)_2(aq) + 2HCl(aq) \rightarrow CaCl_2(aq) + 2H_2O(l)$

    b. $HNO_3(aq) + NaOH(aq) \rightarrow H_2O(l) + NaNO_3(aq)$

    c. $FeS(s) + H_2SO_4(aq) \rightarrow H_2S(g) + FeSO_4(aq)$

    d. $BaCl_2(aq) + K_2CO_3(aq) \rightarrow BaCO_3(s) + 2KCl(aq)$

11 The following answers show the original reaction, the expanded reaction, the expanded reaction with spectator ions crossed out, and the final balanced reaction.

    a. $LiOH(aq) + HI(aq) \rightarrow H_2O(l) + LiI(aq)$

    $Li^+(aq) + OH^-(aq) + H^+(aq) + I^-(aq) \rightarrow H_2O(l) + Li^+(aq) + I^-(aq)$

    $\cancel{Li^+(aq)} + OH^-(aq) + H^+(aq) + \cancel{I^-(aq)} \rightarrow H_2O(l) + \cancel{Li^+(aq)} + \cancel{I^-(aq)}$

    $\mathbf{OH^-(aq) + H^+(aq) \rightarrow H_2O(l)}$

    b. $AgNO_3(aq) + NaCl(aq) \rightarrow AgCl(s) + NaNO_3(aq)$

    $Ag^+(aq) + \cancel{NO_3^-(aq)} + \cancel{Na^+(aq)} + Cl^-(aq) \rightarrow AgCl(s) + \cancel{Na^+(aq)} + \cancel{NO_3^-(aq)}$

    $Ag^+(aq) + \cancel{NO_3^-(aq)} + \cancel{Na^+(aq)} + Cl^-(aq) \rightarrow AgCl(s) + \cancel{Na^+(aq)} + \cancel{NO_3^-(aq)}$

    $\mathbf{Ag^+(aq) + Cl^-(aq) \rightarrow AgCl(s)}$

    c. $Pb(NO_3)_2(aq) + H_2SO_4(aq) \rightarrow PbSO_4(s) + HNO_3(aq)$

    $Pb^{2+}(aq) + 2NO_3^-(aq) + 2H^+(aq) + SO_4^{2-}(aq) \rightarrow PbSO_4(s) + H^+(aq) + NO_3^-(aq)$

    $Pb^{2+}(aq) + \cancel{2NO_3^-(aq)} + \cancel{2H^+(aq)} + SO_4^{2-}(aq) \rightarrow PbSO_4(s) + \cancel{H^+(aq)} + \cancel{NO_3^-(aq)}$

    $\mathbf{Pb^{2+}(aq) + SO_4^{2-}(aq) \rightarrow PbSO_4(s)}$

12   Here are the answers for the second batch of net ionic equation questions, again showing the original reaction, the expanded reaction, the expanded reaction with spectator ions crossed out, and the final balanced reaction.

a. $HCl(aq) + ZnS(aq) \rightarrow H_2S(g) + ZnCl_2(aq)$

$H^+(aq) + Cl^-(aq) + Zn^{2+}(aq) + S^{2-}(aq) \rightarrow H_2S(g) + Zn^{2+}(aq) + 2Cl^-(aq)$

$H^+(aq) + \cancel{Cl^-(aq)} + \cancel{Zn^{2+}(aq)} + S^{2-}(aq) \rightarrow H_2S(g) + \cancel{Zn^{2+}(aq)} + \cancel{2Cl^-(aq)}$

$\mathbf{2H^+(aq) + S^{2-}(aq) \rightarrow H_2S(g)}$

b. $Ca(OH)_2(aq) + H_3PO_4(aq) \rightarrow Ca_3(PO_4)_2(aq) + H_2O(l)$

$Ca^{2+}(aq) + 2OH^-(aq) + 3H^+(aq) + PO_4^{3-}(aq) \rightarrow H_2O(l) + 3Ca^{2+}(aq) + 2PO_4^{3-}(aq)$

$\cancel{Ca^{2+}(aq)} + 2OH^-(aq) + 3H^+(aq) + \cancel{PO_4^{3-}(aq)} \rightarrow H_2O(l) + \cancel{3Ca^{2+}(aq)} + \cancel{2PO_4^{3-}(aq)}$

$\mathbf{OH^-(aq) + H^+(aq) \rightarrow H_2O(l)}$

c. $(NH_4)_2S(aq) + Co(NO_3)_2(aq) \rightarrow CoS(s) + NH_4NO_3(aq)$

$2NH_4^+(aq) + S^{2-}(aq) + Co^{2+}(aq) + 2NO_3^-(aq) \rightarrow CoS(s) + NH_4^+(aq) + NO_3^-(aq)$

$\cancel{2NH_4^+(aq)} + S^{2-}(aq) + Co^{2+}(aq) + \cancel{2NO_3^-(aq)} \rightarrow CoS(s) + \cancel{NH_4^+(aq)} + \cancel{NO_3^-(aq)}$

$\mathbf{S^{2-}(aq) + Co^{2+}(aq) \rightarrow CoS(s)}$

Chapter 9

Putting Stoichiometry to Work

. . . .

In This Chapter

▶ Doing conversions: mole-mole, mole-particles, mole-volume, and mole-mass

▶ Figuring out what happens when one reagent runs out before the others

▶ Using percent yield to determine the efficiency of reactions

. . . .

*S*toichiometry. Such a complicated word for such a simple idea. The Greek roots of the word mean "measuring elements," which doesn't sound nearly as intimidating. Moreover, the ancient Greeks couldn't tell an ionic bond from an ionic column, so just how technical and scary could stoichiometry really be? Simply stated, stoichiometry (pronounced stoh-ick-eee-*ah*-muh-tree) is the quantitative relationship between components of chemical substances. In compound formulas and reaction equations, you express stoichiometry by using subscripted numbers and coefficients.

If you arrived in this chapter by first wandering through Chapters 6, 7, and 8, then you've already had breakfast, lunch, and an afternoon snack with stoichiometry. If you bypassed the aforementioned chapters, then you haven't eaten all day. Either way, it's time for dinner. Please pass the coefficients.

Using Mole-Mole Conversions from Balanced Equations

Mass and energy are conserved. It's the Law. Unfortunately, this means that there's no such thing as a free lunch, or any other type of free meal. Ever. On the other hand, the conservation of mass makes it possible to predict how chemical reactions will turn out.

Chapter 8 describes why chemical reaction equations should be balanced for equal mass in reactants and products. You balance an equation by adjusting the coefficients that precede reactant and product compounds within the equation. Balancing equations can seem like a chore, like taking out the trash. But a balanced equation is far better than any collection of coffee grounds and orange peels because such an equation is a useful tool. After you've got a balanced equation, you can use the coefficients to build *mole-mole conversion factors*. These kinds of conversion factors tell you how much of any given product you get by reacting any given amount of reactant. This is one of those calculations that makes chemists particularly useful, so they needn't get by on looks and charm alone.

Consider the following balanced equation for generating ammonia from nitrogen and hydrogen gases:

$$N_2(g) + 3H_2(g) \rightarrow 2NH_3(g)$$

Industrial chemists around the globe perform this reaction, humorlessly fixating on how much ammonia product they'll end up with at the end of the day. (In fact, clever methods for improving the rate and yield of this reaction garnered Nobel Prizes for two German gentlemen, die Herren Haber und Bosch.) In any event, how are chemists to judge how closely their reactions have approached completion? The heart of the answer lies in a balanced equation and the mole-mole conversion factors that spring from it.

For every mole of dinitrogen reactant, a chemist expects 2 moles of ammonia product. Similarly, for every 3 moles of dihydrogen reactant, the chemist expects 2 moles of ammonia product. These expectations are based on the coefficients of the balanced equation and are expressed as mole-mole conversion factors as shown in Figure 9-1.

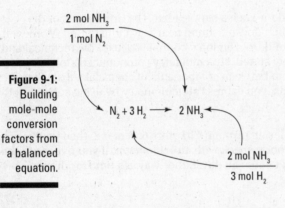

Figure 9-1:
Building
mole-mole
conversion
factors from
a balanced
equation.

Q. How many moles of ammonia can be expected from the reaction of 278 moles of N_2 gas?

A. **556 moles of ammonia.** Begin with your known quantity, the 278 moles of dinitrogen that is to be reacted. Multiply that quantity by the mole-mole conversion factor that relates moles of dinitrogen to moles of ammonia. Use the orientation of the conversion factor that places mol NH_3 on top and mol N_2 on the bottom. This way, the mol N_2 units cancel, leaving you with the desired units, mol NH_3:

$$\frac{278 \text{ mol } N_2}{1} \times \frac{2 \text{ mol } NH_3}{1 \text{ mol } N_2} = 556 \text{ mol } NH_3$$

1. One source of dihydrogen gas is from the electrolysis of water, in which electricity is passed through water to break hydrogen-oxygen bonds, yielding dihydrogen and dioxygen gases:

$$2H_2O(l) \rightarrow 2H_2(g) + O_2(g)$$

a. How many moles of dihydrogen gas result from the electrolysis of 78.4 moles of water?

b. How many moles of water are required to produce 905 moles of dihydrogen?

c. Running the electrolysis reaction in reverse constitutes the combustion of hydrogen. How many moles of dioxygen are required to combust 84.6 moles of dihydrogen?

Solve It

2. Aluminum reacts with copper (II) sulfate to produce aluminum sulfate and copper, as summarized by the skeleton equation:

$$Al(s) + CuSO_4(aq) \rightarrow Al_2(SO_4)_3(aq) + Cu(s)$$

a. Balance the equation.

b. How many moles of aluminum are needed to react with 10.38 moles of copper (II) sulfate?

c. How many moles of copper are produced if 2.08 moles of copper (II) sulfate react with aluminum?

d. How many moles of copper (II) sulfate are needed to produce 0.96 moles of aluminum sulfate?

e. How many moles of aluminum are needed to produce 20.01 moles of copper?

Solve It

3. Solid iron reacts with solid sulfur to form iron (III) sulfide:

$$Fe(s) + S(s) \rightarrow Fe_2S_3(s)$$

a. Balance the equation.

b. With how many moles of sulfur do 6.2 moles of iron react?

c. How many moles of iron (III) sulfide are produced from 10.6 moles of iron?

d. How many moles of iron (III) sulfide are produced from 3.5 moles of sulfur?

Solve It

4. Ethane combusts to form carbon dioxide and water:

$$C_2H_6(g) + O_2(g) \rightarrow CO_2(g) + H_2O(g)$$

a. Balance the equation.

b. 15.4 moles of ethane produces how many moles of carbon dioxide?

c. How many moles of ethane does it take to produce 293 moles of water?

d. How many moles of oxygen are required to combust 0.178 moles of ethane?

Solve It

Putting Moles at the Center: Conversions Involving Particles, Volumes, and Masses

The mole is the beating heart of stoichiometry, the central unit through which other quantities flow. Real-life chemists don't have magic mole vision, however. A chemist can't look at a pile of potassium chloride crystals, squint her eyes, and proclaim: "That's 0.539 moles of salt." Well, she could proclaim such a thing, but she wouldn't bet her pocket protector on it. Real *reagents* (reactants) tend to be measured in units of mass or volume, and occasionally even in actual numbers of particles. Real products are measured in the same way. So, you need to be able to use *mole-mass, mole-volume,* and *mole-particle conversion factors* to translate between these different dialects of counting. Figure 9-2 summarizes the interrelationship between all these things and serves as a flowchart for problem solving. All roads lead to and from the mole.

Figure 9-2: A problem-solving flowchart showing the use of mole-mole, mole-mass, mole-volume, and mole-particle conversion factors.

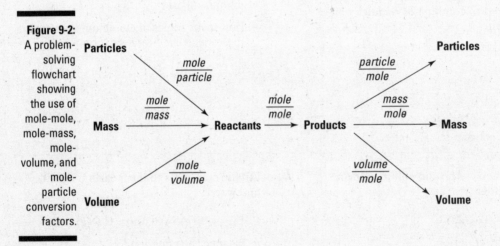

Q. Calcium carbonate decomposes to produce solid calcium oxide and carbon dioxide gas according to the following reaction. Answer each part of the question assuming that 10.0g of calcium carbonate decomposes.

$CaCO_3(s) \rightarrow CaO(s) + CO_2(g)$

a. How many grams of calcium oxide are produced?

b. At standard temperature and pressure (STP), how many liters of carbon dioxide are produced?

A. **a. 5.61g CaO.** First, convert 10.0g of calcium carbonate to moles of calcium carbonate by using the molar mass of calcium carbonate ($100g \, mol^{-1}$) as a conversion factor. This step precedes the others in the calculation for each question. To determine the grams of calcium oxide produced, follow the initial conversion with a mole-mole conversion to find the moles of calcium oxide produced. Then convert from moles of calcium oxide to grams of calcium oxide by using the molar mass of calcium oxide ($56.1g \ mol^{-1}$) as a conversion factor:

$$\frac{10.0g \ CaCO_3}{1} \times \frac{1 \ mol \ CaCO_3}{100g \ CaCO_3} \times \frac{1 \ mol \ CaO}{1 \ mol \ CaCO_3} \times \frac{56.1g \ CaO}{1 \ mol \ CaO} = 5.61g \ CaO$$

b. 2.24L CO_2. To determine the liters of carbon dioxide produced, follow the initial mass-mole conversion with a mole-mole conversion to find the moles of carbon dioxide produced. Then convert from moles of carbon dioxide to liters by using the assumption that, at STP, each mole of gas occupies 22.4L (see Chapter 7 for details):

$$\frac{10.0g \ CaCO_3}{1} \times \frac{1 \ mol \ CaCO_3}{100g \ CaCO_3} \times \frac{1 \ mol \ CO_2}{1 \ mol \ CaCO_3} \times \frac{22.4L \ CO_2}{1 \ mol \ CO_2} = 2.24L \ CO_2$$

Notice how both calculations require you first to translate into the language of moles, and then perform a mole-mole conversion using stoichiometry from the reaction equation. Finally, you convert into the desired units. Both solutions consist of a chain of conversion factors, each factor bringing the units one step closer to those needed in the answer.

---

5. Hydrogen peroxide decomposes into oxygen gas and liquid water:

$$H_2O_2(l) \rightarrow H_2O(l) + O_2(g)$$

a. Balance the equation.

b. How many grams of water are produced when 2.94×10^{24} molecules of hydrogen peroxide decompose?

c. What is the volume of oxygen produced (at STP) when 32.9g of hydrogen peroxide decompose?

Solve It

6. Dinitrogen trioxide gas reacts with liquid water to produce nitrous acid:

$$N_2O_3(g) + H_2O(l) \rightarrow HNO_2(aq)$$

a. Balance the equation.

b. At STP, how many liters of dinitrogen trioxide react to produce 36.98g of dissolved nitrous acid?

c. How many molecules of water react with 17.3L of dinitrogen trioxide (at STP)?

Solve It

7. Calcium phosphate and liquid water are produced from the reaction of calcium hydroxide with phosphoric acid:

$Ca(OH)_2(aq) + H_3PO_4(aq) \rightarrow Ca_3(PO_4)_2(aq) + H_2O(l)$

a. Balance the equation.

b. 62.8g of calcium hydroxide are dissolved into a solution of phosphoric acid. All the calcium hydroxide reacts. How many molecules of water are produced?

c. Phosphoric acid reacts to produce 133g of calcium phosphate. How many grams of dissolved calcium hydroxide reacted?

Solve It

8. Lead (II) chloride reacts with chlorine to produce lead (IV) chloride:

$PbCl_2(s) + Cl_2(g) \rightarrow PbCl_4(l)$

a. What volume of chlorine reacts to convert 50.0g of lead (II) chloride entirely into product?

b. How many *formula units* (that is, moles of ionic compound) of lead (IV) chloride result from the reaction of 13.71g of lead (II) chloride?

c. How many grams of lead (II) chloride react to produce 84.8g of lead (IV) chloride?

Solve It

Limiting Your Reagents

In real-life chemical reactions, not all of the reactants present convert into product. That would be perfect and convenient. Does that sound like real life to you? More typically, one reagent is completely used up, and others are left in excess, perhaps to react another day.

The situation resembles that of a horde of Hollywood hopefuls lined up for a limited number of slots as extras in a film. Only so many eager faces react with an available slot to produce a happily (albeit pitifully) employed actor. The remaining actors are in excess, muttering quietly all the way back to their jobs as waiters. In this scenario, the slots are the limiting reagent.

Those standing in line demand to know, how many slots are there? With this key piece of data, they can deduce how many of their huddled mass will end up with a gig. Or, they can figure out how many will continue to waste their film school degrees serving penne with basil and goat cheese to chemists on vacation.

 Chemists demand to know, which reactant will run out first? In other words, which reactant is the *limiting reagent?* Knowing that information allows them to deduce how much product they can expect, based on how much of the limiting reagent they've put into the reaction. Also, identifying the limiting reagent allows them to calculate how much of the excess reagent they'll have left over when all the smoke clears. Either way, the first step is to figure out which is the limiting reagent.

In any chemical reaction, you can simply pick one reagent as a candidate for the limiting reagent, calculate how many moles of that reagent you have, and then calculate how many grams of the other reagent you'd need to react both to completion. You'll discover one of two things. Either you have an excess of the first reagent, or you have an excess of the second reagent. The one you have in excess is the *excess reagent*. The one that isn't in excess is the limiting reagent.

Q. Ammonia reacts with oxygen to produce nitrogen monoxide and liquid water:

$NH_3(g) + O_2(g) \rightarrow NO(g) + H_2O(l)$

a. Balance the equation.

b. Determine the limiting reagent if 100g of ammonia and 100g of oxygen are present at the beginning of the reaction.

c. What is the excess reagent and how many grams of the excess reagent will remain when the reaction reaches completion?

d. How many grams of nitrogen monoxide will be produced if the reaction goes to completion?

e. How many grams of water will be produced if the reaction goes to completion?

A. **a.** Before doing anything else, you must have a balanced reaction equation. Don't waste good thought on an unbalanced equation. The balanced form of the given equation is:

$4NH_3(g) + 5O_2(g) \rightarrow 4NO(g) + 6H_2O(l)$

b. Oxygen is the limiting reagent. Two candidates vie for the status of limiting reagent, NH_3 and O_2. You start with 100g of each, which corresponds to some number of moles of each. Furthermore, you require 4 moles of ammonia for every 5 moles of oxygen gas.

Starting with ammonia, the calculation to determine the limiting reagent goes as shown in the following equation:

$$\frac{100g \ NH_3}{1} \times \frac{1 \ mol \ NH_3}{17.0g \ NH_3} \times \frac{5 \ mol \ O_2}{4 \ mol \ NH_3} \times \frac{32.0g \ O_2}{1 \ mol \ O_2} = 235g \ O_2$$

The calculation reveals that you'd need 235g of oxygen gas to completely react with 100g of ammonia. But you only have 100g of oxygen. You'll run out of oxygen before you run out of ammonia, so oxygen is the limiting reagent.

c. The excess reagent is ammonia, and 57.5g of ammonia will remain when the reaction reaches completion. To calculate how many grams of ammonia will be left at the end of the reaction, assume that all 100g of oxygen react:

$$\frac{100g \ O_2}{1} \times \frac{1 \ mol \ O_2}{32.0g \ O_2} \times \frac{4 \ mol \ NH_3}{5 \ mol \ O_2} \times \frac{17.0g \ NH_3}{1 \ mol \ NH_3} = 42.5g \ NH_3$$

This calculation shows that 42.5g of the original 100g of ammonia will react before the limiting reagent is expended. So, 100g – 42.5g = 57.5g of ammonia will remain in excess.

d. 75g of nitrogen monoxide will be produced. To determine the grams of nitrogen monoxide that are generated by the complete reaction of oxygen, start with the assumption that all 100g of the oxygen react:

$$\frac{100g\ O_2}{1} \times \frac{1\ mol\ O_2}{32.0g\ O_2} \times \frac{4\ mol\ NO}{5\ mol\ O_2} \times \frac{30.0g\ NO}{1\ mol\ NO} = 75.0g\ NO$$

e. 67.5g of water will be produced. Again, assume that all 100g of the oxygen react in order to determine how many grams of water are produced:

$$\frac{100g\ O_2}{1} \times \frac{1\ mol\ O_2}{32.0g\ O_2} \times \frac{6\ mol\ H_2O}{5\ mol\ O_2} \times \frac{18.0g\ H_2O}{1\ mol\ H_2O} = 67.5g\ H_2O$$

Reassuringly, the total mass of products (75g + 67.5g = 142.5g) equals the total mass of reactants (100g + 42.5g = 142.5g), so the Conservation-of-Mass Police needn't strap on the riot gear.

---

9. Iron (III) oxide reacts with carbon monoxide to produce iron and carbon dioxide:

$Fe_2O_3(s) + 3CO(g) \rightarrow 2Fe(s) + 3CO_2(g)$

a. What is the limiting reagent if 50g of iron (III) oxide and 67g of carbon monoxide are present at the beginning of the reaction?

b. What is the excess reagent, and how many grams of it will remain after the reaction proceeds to completion?

c. How many grams of each product should be expected if the reaction goes to completion?

Solve It

10. Solid sodium reacts (violently) with water to produce sodium hydroxide and hydrogen gas:

$2Na(s) + 2H_2O(l) \rightarrow 2NaOH(aq) + H_2(g)$

a. What is the limiting reagent if 25g of sodium and 40.2g of water are present at the beginning of the reaction?

b. What is the excess reagent, and how many grams of it will remain after the reaction has gone to completion?

c. How many grams of sodium hydroxide and how many liters (at STP) of hydrogen gas should be expected if the reaction goes to completion?

Solve It

11. Aluminum reacts with chlorine gas to produce aluminum chloride:

$$2Al(s) + 3Cl_2(g) \rightarrow 2AlCl_3(s)$$

a. What is the limiting reagent if 29.3g of aluminum and 34.6L (at STP) of chlorine gas are present at the beginning of the reaction?

b. What is the excess reagent, and how many grams (or liters at STP) of it will remain after the reaction has gone to completion?

c. How many grams of aluminum chloride should be expected if the reaction goes to completion?

Solve It

Counting Your Chickens after They've Hatched: Percent Yield Calculations

In a way, reactants have it easy. Maybe they'll make something of themselves and actually react. Or maybe they'll just lean against the inside of the beaker, flipping through a back issue of *People* magazine and sipping a caramel macchiato.

Chemists don't have it so easy. Someone is paying them to do reactions. That someone doesn't have time or money for excuses about loitering reactants. So you, as a fresh-faced chemist, have to be concerned with just how completely your reactants react to form products. To compare the amount of product obtained from a reaction with the amount that should have been obtained, chemists use *percent yield*. You determine percent yield with the following formula:

Percent yield = 100% × (Actual yield) / (Theoretical yield)

Lovely, but what is an actual yield and what is a theoretical yield? An *actual yield* is . . . well . . . the amount of product actually produced by the reaction. A *theoretical yield* is the amount of product that could have been produced had everything gone perfectly, as described by theory — in other words, as predicted by your painstaking calculations.

Things never go perfectly. Reagents stick to the sides of flasks. Impurities sabotage reactions. Chemists attempt to dance. None of these ghastly things are *supposed* to occur, but they do. So, actual yields fall short of theoretical yields.

Q. Calculate the percent yield of sodium sulfate in the following scenario: 32.18g of sulfuric acid reacts with excess sodium hydroxide to produce 37.91g of sodium sulfate.

$$H_2SO_4(aq) + 2NaOH(aq) \rightarrow 2H_2O(l) + Na_2SO_4(aq)$$

A. **81.37% is the percent yield.** The question makes clear that sodium hydroxide is the excess reagent. So, sulfuric acid is the limiting reagent and is the reagent you should use to calculate the theoretical yield:

$$\frac{32.18g\ H_2SO_4}{1} \times \frac{1\ mol\ H_2SO_4}{98.08g\ H_2SO_4} \times \frac{1\ mol\ Na_2SO_4}{1\ mol\ H_2SO_4} \times \frac{142.0g\ Na_2SO_4}{1\ mol\ Na_2SO_4} = 46.59g\ Na_2SO_4$$

Theory predicts that 46.59g of sodium sulfate product are possible if the reaction proceeds perfectly and to completion. But the question states that the actual yield was only 37.91g of sodium sulfate. With these two pieces of information, you can calculate the percent yield:

Percent yield = $100\% \times (37.91g) / (46.59g) = 81.37\%$

---

12. Sulfur dioxide reacts with liquid water to produce sulfurous acid:

$$SO_2(g) + H_2O(l) \rightarrow H_2SO_3(aq)$$

a. What is the percent yield if 19.07g of sulfur dioxide reacts with excess water to produce 21.61g of sulfurous acid?

b. When 8.11g of water reacts with excess sulfur dioxide, 27.59g of sulfurous acid is produced. What is the percent yield?

Solve It

13. Liquid hydrazine is a component of some rocket fuels. Hydrazine combusts to produce nitrogen gas and water:

$$N_2H_4(l) + O_2(g) \rightarrow N_2(g) + 2H_2O(g)$$

a. If the percent yield of a combustion reaction in the presence of 23.4g of N_2H_4 (and excess oxygen) is 98%, how many liters of dinitrogen (at STP) are produced?

b. What is the percent yield if 84.8g of N_2H_4 reacts with 54.7g of oxygen gas to produce 51.33g of water?

Solve It

Answers to Questions on Stoichiometry

Balanced equations, diverse conversion factors, limiting reagents, and percent yield calculations all now tremble before you. Probably. Be sure your powers over them are as breathtaking as they should be by checking your answers.

1 The answers to parts a, b, and c are

a. 78.4 mol H$_2$.

$$\frac{78.4 \text{ mol H}_2\text{O}}{1} \times \frac{2 \text{ mol H}_2}{2 \text{ mol H}_2\text{O}} = 78.4 \text{ mol H}_2$$

b. 905 mol H$_2$O.

$$\frac{905 \text{ mol H}_2}{1} \times \frac{2 \text{ mol H}_2\text{O}}{2 \text{ mol H}_2} = 905 \text{ mol H}_2\text{O}$$

c. 42.3 mol O$_2$.

$$\frac{84.6 \text{ mol H}_2}{1} \times \frac{1 \text{ mol O}_2}{2 \text{ mol H}_2} = 42.3 \text{ mol O}_2$$

2 Before attempting to do any calculations, you must balance the equation as shown in part a (see Chapter 8 for more information).

a. 2Al(s) + 3CuSO$_4$(aq) → Al$_2$(SO$_4$)$_3$(aq) + 3Cu(s)

b. 6.920 mol Al.

$$\frac{10.38 \text{ mol CuSO}_4}{1} \times \frac{2 \text{ mol Al}}{3 \text{ mol CuSO}_4} = 6.920 \text{ mol Al}$$

c. 2.08 mol Cu.

$$\frac{2.08 \text{ mol CuSO}_4}{1} \times \frac{3 \text{ mol Cu}}{3 \text{ mol CuSO}_4} = 2.08 \text{ mol Cu}$$

d. 2.9 mol CuSO$_4$.

$$\frac{0.96 \text{ mol Al}_2\left(\text{SO}_4\right)_3}{1} \times \frac{3 \text{ mol CuSO}_4}{1 \text{ mol Al}_2\left(\text{SO}_4\right)_3} = 2.9 \text{ mol CuSO}_4$$

e. 13.34 mol Al.

$$\frac{20.01 \text{ mol Cu}}{1} \times \frac{2 \text{ mol Al}}{3 \text{ mol Cu}} = 13.34 \text{ mol Al}$$

3 Balance the equation first, and then go on to the calculations for the remaining questions.

a. 2Fe(s) + 3S(s) → Fe$_2$S$_3$(s)

b. 9.3 mol S.

$$\frac{6.2 \text{ mol Fe}}{1} \times \frac{3 \text{ mol S}}{2 \text{ mol Fe}} = 9.3 \text{ mol S}$$

c. 5.30 mol Fe₂S₃.

$$\frac{10.6 \text{ mol Fe}}{1} \times \frac{1 \text{ mol Fe}_2S_3}{2 \text{ mol Fe}} = 5.30 \text{ mol Fe}_2S_3$$

d. 1.2 mol Fe₂S₃.

$$\frac{3.5 \text{ mol S}}{1} \times \frac{1 \text{ mol Fe}_2S_3}{3 \text{ mol S}} = 1.2 \text{ mol Fe}_2S_3$$

4 As always, begin by balancing the equation.

a. $2C_2H_6(g) + 7O_2(g) \rightarrow 4CO_2(g) + 6H_2O(g)$

b. 30.8 mol CO₂.

$$\frac{15.4 \text{ mol C}_2H_6}{1} \times \frac{4 \text{ mol CO}_2}{2 \text{ mol C}_2H_6} = 30.8 \text{ mol CO}_2$$

c. 97.7 mol C₂H₆.

$$\frac{293 \text{ mol H}_2O}{1} \times \frac{2 \text{ mol C}_2H_6}{6 \text{ mol H}_2O} = 97.7 \text{ mol C}_2H_6$$

d. 0.623 mol O₂.

$$\frac{0.178 \text{ mol C}_2H_6}{1} \times \frac{7 \text{ mol O}_2}{2 \text{ mol C}_2H_6} = 0.623 \text{ mol O}_2$$

5 Without a balanced equation, nothing else is possible.

a. $2H_2O_2(l) \rightarrow 2H_2O(l) + O_2(g)$

b. 87.9g H₂O.

$$\frac{2.94 \times 10^{24} \text{ molec H}_2O_2}{1} \times \frac{1 \text{ mol}}{6.022 \times 10^{23} \text{molec}} \times \frac{2 \text{ mol H}_2O}{2 \text{ mol H}_2O_2} \times \frac{18.0g \text{ H}_2O}{1 \text{ mol H}_2O} = 87.9g \text{ H}_2O$$

c. 10.8L O₂.

$$\frac{32.9g \text{ H}_2O_2}{1} \times \frac{1 \text{ mol H}_2O_2}{34.0g \text{ H}_2O_2} \times \frac{1 \text{ mol O}_2}{2 \text{ mol H}_2O_2} \times \frac{22.4L \text{ O}_2}{1 \text{ mol O}_2} = 10.8L \text{ O}_2$$

6 The conversion factors you need to do the calculations require a balanced equation, so do that first.

a. $N_2O_3(g) + H_2O(l) \rightarrow 2HNO_2(aq)$

b. 8.81L N₂O₃.

$$\frac{36.98g \text{ HNO}_2}{1} \times \frac{1 \text{ mol HNO}_2}{47.0g \text{ HNO}_2} \times \frac{1 \text{ mol N}_2O_3}{2 \text{ mol HNO}_2} \times \frac{22.4L \text{ N}_2O_3}{1 \text{ mol N}_2O_3} = 8.81L \text{ N}_2O_3$$

c. 4.65 × 10²³ molec H₂O.

$$\frac{17.3L \text{ N}_2O_3}{1} \times \frac{1 \text{ mol N}_2O_3}{22.4L \text{ N}_2O_3} \times \frac{1 \text{ mol H}_2O}{1 \text{ mol N}_2O_3} \times \frac{6.022 \times 10^{23} \text{ molec H}_2O}{1 \text{ mol H}_2O} = 4.65 \times 10^{23} \text{ molec H}_2O$$

7 That's right — balancing the equation should be your first priority.

a. $3Ca(OH)_2(aq) + 2H_3PO_4(aq) \rightarrow Ca_3(PO_4)_2(aq) + 6H_2O(l)$

b. 1.02×10^{24} **molec H_2O.**

$$\frac{62.8g \, Ca(OH)_2}{1} \times \frac{1 \, mol \, Ca(OH)_2}{74.1g \, Ca(OH)_2} \times \frac{6 \, mol \, H_2O}{3 \, mol \, Ca(OH)_2} \times \frac{6.022 \times 10^{23} \, molec \, H_2O}{1 \, mol \, H_2O} = 1.02 \times 10^{24} \, molec \, H_2O$$

c. 95.3g Ca(OH)$_2$.

$$\frac{133g \, Ca_3(PO_4)_2}{1} \times \frac{1 \, mol \, Ca_3(PO_4)_2}{310.2g \, Ca_3(PO_4)_2} \times \frac{3 \, mol \, Ca(OH)_2}{1 \, mol \, Ca_3(PO_4)_2} \times \frac{74.1g \, Ca(OH)_2}{1 \, mol \, Ca(OH)_2} = 95.3g \, Ca(OH)_2$$

8 In this problem, the provided chemical equation is already balanced, so you can proceed directly to the calculations for parts a through c:

a. 4.03L Cl$_2$.

$$\frac{50.0g \, PbCl_2}{1} \times \frac{1 \, mol \, PbCl_2}{278.1g \, PbCl_2} \times \frac{1 \, mol \, Cl_2}{1 \, mol \, PbCl_2} \times \frac{22.4L \, Cl_2}{1 \, mol \, Cl_2} = 4.03L \, Cl_2$$

b. 2.97×10^{22} form. un. PbCl$_4$.

$$\frac{13.71g \, PbCl_2}{1} \times \frac{1 \, mol \, PbCl_2}{278.1g \, mol \, PbCl_2} \times \frac{1 \, mol \, PbCl_4}{1 \, mol \, PbCl_2} \times$$

$$\frac{6.022 \times 10^{23} \, form. \, un. \, PbCl4}{1 \, mol \, PbCl4} = 2.97 \times 10^{22} \, form. \, un. \, PbCl4$$

c. 67.6g PbCl$_2$.

$$\frac{84.8g \, PbC14}{1} \times \frac{1 \, mol \, PbCl4}{349.0g \, mol \, PbCl4} \times \frac{1 \, mol \, PbCl2}{1 \, mol \, PbCl4} \times \frac{278.1g \, PbCl2}{1 \, mol \, PbCl2} = 67.6g \, PbCl2$$

9 Begin limiting reagent problems by determining which is the limiting reagent. The answers to other questions build on that foundation. To find the limiting reagent, simply pick one of the reactants as a candidate. How much of your candidate reactant do you have? Use that information to calculate how much of any other reactants you'd need for a complete reaction. You can deduce the limiting and excess reagents from the results of these calculations.

a. Iron (III) oxide. In this example, iron (III) oxide was chosen as the initial candidate limiting reagent. It turned out to be the correct choice because more carbon monoxide is initially present than is required to react with all the iron (III) oxide.

$$\frac{50g \, Fe_2O_3}{1} \times \frac{1 \, mol \, Fe_2O_3}{159.7g \, Fe_2O_3} \times \frac{3 \, mol \, CO}{1 \, mol \, Fe_2O_3} \times \frac{28.0g \, CO}{1 \, mol \, CO} = 26g \, CO$$

b. **Carbon monoxide is the excess reagent, and 41g of it will remain after the reaction is completed.** Because iron (III) oxide is the limiting reagent, the excess reagent is carbon monoxide.

To find the amount of excess reagent that will remain after the reaction reaches completion, first calculate how much of the excess reagent will be consumed. This calculation is identical to the one performed in part a. So, 26g of carbon monoxide will be consumed. Next, subtract the quantity consumed from the amount originally present to obtain the amount of carbon monoxide that will remain: 67g – 26g = 41g.

c. **35g of iron; 41g of carbon dioxide.** To answer this question, do two calculations, each starting with the assumption that all the limiting reagent is consumed:

$$\frac{50g\ Fe_2O_3}{1} \times \frac{1\ mol\ Fe_2O_3}{159.7g\ Fe_2O_3} \times \frac{2\ mol\ Fe}{1\ mol\ Fe_2O_3} \times \frac{55.85g\ Fe}{1\ mol\ Fe} = 35g\ Fe$$

$$\frac{50g\ Fe_2O_3}{1} \times \frac{1\ mol\ Fe_2O_3}{159.7g\ Fe_2O_3} \times \frac{3\ mol\ CO_2}{1\ mol\ Fe_2O_3} \times \frac{44.0g\ CO_2}{1\ mol\ CO_2} = 41g\ CO_2$$

10 To find the limiting reagent, simply pick one of the reactants as a candidate. Calculate how many of the other reagents you'd need to completely react with all of your available candidate reagent. Deduce the limiting and excess reagents from these calculations.

a. **Sodium.** In this example, sodium was chosen as the initial candidate limiting reagent. It turned out to be the correct choice because more water is initially present than is required to react with all the sodium.

$$\frac{25g\ Na}{1} \times \frac{1\ mol\ Na}{23.0g\ Na} \times \frac{2\ mol\ H_2O}{2\ mol\ Na} \times \frac{18.0g\ H_2O}{1\ mol\ H_2O} = 20g\ H_2O$$

b. **Water is the excess reagent, and 20.2g of it will remain after the reaction is completed.** Sodium is the limiting reagent, so the excess reagent is water.

The calculation in part a revealed how much water is consumed in a complete reaction: 20g. Because 40.2g of water are initially present, 20.2g of water will remain after the reaction: 40.2g – 20g = 20.2g.

c. **43g of sodium hydroxide; 12L of hydrogen gas.** To answer this question, do two calculations, each starting with the assumption that all the limiting reagent is consumed:

$$\frac{25g\ Na}{1} \times \frac{1\ mol\ Na}{23.0g\ Na} \times \frac{2\ mol\ NaOH}{2\ mol\ Na} \times \frac{40g\ NaOH}{1\ mol\ NaOH} = 43g\ NaOH$$

$$\frac{25g\ Na}{1} \times \frac{1\ mol\ Na}{23.0g\ Na} \times \frac{1\ mol\ H_2}{2\ mol\ Na} \times \frac{22.4L\ H_2}{1\ mol\ H_2} = 12L\ H_2$$

11 The first step is to identify the limiting reagent. Simply pick one of the reactants as a trial candidate for the limiting reagent, and calculate how much of the other reagents are required to react completely with the candidate.

a. **Chlorine gas.** In this example, aluminum was chosen as the candidate limiting reagent. It turns out that more chlorine gas is required (36.5L) to completely react with the available aluminum than is available (34.6L). So, chlorine gas is the limiting reagent.

$$\frac{29.3g\ Al}{1} \times \frac{1\ mol\ Al}{27.0g\ Al} \times \frac{3\ mol\ Cl_2}{2\ mol\ Al} \times \frac{22.4L\ Cl_2}{1\ mol\ Cl_2} = 36.5L\ Cl_2$$

b. Aluminum is the excess reagent, and 1.5g of it will remain after the reaction is completed. To calculate how much excess reagent (aluminum, in this case) will remain after a complete reaction, first calculate how much will be consumed, as shown in the equation that follows. It turns out that 27.8g of aluminum will be consumed. Subtract that quantity from the amount originally present to calculate the remaining amount: 29.3g – 27.8g = 1.5g.

$$\frac{34.6\text{L Cl}_2}{1} \times \frac{1\text{ mol Cl}_2}{22.4\text{L Cl}_2} \times \frac{2\text{ mol Al}}{3\text{ mol Cl}_2} \times \frac{27.0\text{g Al}}{1\text{ mol Al}} = 27.8\text{g Al}$$

c. 137g of aluminum chloride. This calculation starts with the assumption that all the limiting reagent, chlorine gas in this case, is consumed.

$$\frac{34.6\text{L Cl}_2}{1} \times \frac{1\text{ mol Cl}_2}{22.4\text{L Cl}_2} \times \frac{2\text{ mol AlCl}_3}{3\text{ mol Cl}_2} \times \frac{133.3\text{g AlCl}_3}{1\text{ mol AlCl}_3} = 137\text{g AlCl}_3$$

12 The reaction equation is already balanced, so you can proceed with the calculations.

a. 88.46 is the percent yield. You're given an actual yield and asked to calculate the percent yield. To do so, you must calculate the theoretical yield. The question makes clear that sulfur dioxide is the limiting reagent, so begin the calculation of theoretical yield by assuming that all the sulfur dioxide is consumed.

$$\frac{19.07\text{g SO}_2}{1} \times \frac{1\text{ mol SO}_2}{64.05\text{g SO}_2} \times \frac{1\text{ mol H}_2\text{SO}_3}{1\text{ mol SO}_2} \times \frac{82.07\text{g H}_2\text{SO}_3}{1\text{ mol H}_2\text{SO}_3} = 24.43\text{g H}_2\text{SO}_3$$

So, the theoretical yield of sulfurous acid is 24.43g. With this information, you can calculate the percent yield:

Percent yield = 100% × (21.61g / 24.43g) = 88.46%

b. 74.6% is the percent yield. Again, you're asked to calculate a percent yield having been given an actual yield. In this case, the limiting reagent is water, so begin the calculation by assuming that all the water is consumed.

$$\frac{8.11\text{g H}_2\text{O}}{1} \times \frac{1\text{ mol H}_2\text{O}}{18.0\text{g H}_2\text{O}} \times \frac{1\text{ mol H}_2\text{SO}_3}{1\text{ mol H}_2\text{O}} \times \frac{82.07\text{g H}_2\text{SO}_3}{1\text{ mol H}_2\text{SO}_3} = 37.0\text{g H}_2\text{SO}_3$$

So, the theoretical yield of sulfurous acid is 37.0g. Using that information, calculate the percent yield:

Percent yield = 100% × (27.59g / 37.0g) = 74.6%

13 With a balanced equation at your disposal, you can calculate with impunity.

a. 16.1L of dinitrogen are produced. In this question, you're given the percent yield and asked to calculate an actual yield. To do so, you must know the theoretical yield. The question makes clear that hydrazine is the limiting reagent, so calculate the theoretical yield of dinitrogen by assuming that all the hydrazine is consumed.

$$\frac{23.4\text{g N}_2\text{H}_4}{1} \times \frac{1\text{ mol N}_2\text{H}_4}{32.05\text{g N}_2\text{H}_4} \times \frac{1\text{ mol N}_2}{1\text{ mol N}_2\text{H}_4} \times \frac{22.4\text{L N}_2}{1\text{ mol N}_2} = 16.4\text{L N}_2$$

So, the theoretical yield of dinitrogen is 16.4L. Algebraically rearrange the percent yield equation to solve for the actual yield:

Actual yield = (Theoretical yield × Percent yield) / 100%

Substitute for your known values and solve:

Actual yield = (16.4L × 98%) / 100% = 16.1L

b. **83.5% is the percent yield.** In this question, you're given the initial amounts of the reactants and an actual yield. You're asked to calculate a percent yield. To do so, you need to know the theoretical yield. The added wrinkle in this problem (as opposed to the one in part a) is that you don't initially know which reagent is limiting. As with any limiting reagent problem, pick a candidate limiting reagent. In this example, hydrazine is chosen as the initial candidate:

$$\frac{84.8g\ N_2H_4}{1} \times \frac{1\ mol\ N_2H_4}{32.05g\ N_2H_4} \times \frac{1\ mol\ O_2}{1\ mol\ N_2H_4} \times \frac{32.0g\ O_2}{1\ mol\ O_2} = 84.7g\ O_2$$

It turns out that more oxygen is required to react with the available hydrazine than is initially present. So, oxygen is the limiting reagent. Knowing this, calculate a theoretical yield of water by assuming that all the oxygen is consumed.

$$\frac{54.7g\ O_2}{1} \times \frac{1\ mol\ O_2}{32.0g\ O_2} \times \frac{1\ mol\ H_2O}{1\ mol\ O_2} \times \frac{18.0g\ H_2O}{1\ mol\ H_2O} = 61.5g\ H_2O$$

So, the theoretical yield of water is 61.5g. Use this information to calculate the percent yield:

Percent yield = 100% × (51.33g / 61.5g) = 83.5%

Part III
Examining Changes in Terms of Energy

The 5th Wave   By Rich Tennant

IRONICALLY, THE LAST THING PROF. CARUTHERS REMEMBERED WAS EXPLAINING HOW ENERGY IS PRODUCED WHEN MOLECULES COLLIDE.

In this part . . .

Any chemical drama unfolds on a stage, each character following an individual script, driven by particular motivations. In this part, we explore some of the common stages for chemical drama: gaseous and liquid solutions. We describe how and why some chemical dramas are one-act monologues while others are 12-act masterworks; why some race to a conclusion, and others stretch interminably over hours, days, and years. Finally, we examine energy, the muse behind the drama, and how differences in energy determine which characters do what and when.

Chapter 10

Understanding States in Terms of Energy

- -

In This Chapter

▶ Explaining kinetic theory

▶ Moving between phases

▶ Distinguishing differences between solids

- -

*W*hen asked, children often report that solids, liquids, and gases are composed of different kinds of matter. This assumption is understandable given the striking differences in the properties of these three states. Nevertheless, for a given type of matter at a given pressure, the fundamental difference between a solid, a liquid, and a gas actually is the *amount of energy* within the particles of matter. Understanding the states of matter (phases) in terms of energy and pressure helps to explain the different properties of those states and how matter moves between the states. We explain what you need to know in this chapter.

Describing States of Matter with Kinetic Theory

Imagine two pool balls, each glued to either end of a spring. How many different kinds of motion could this contraption undergo? You could twist along the axis of the spring. You could bend the spring or stretch it. You could twirl the whole thing around, or you could throw it through the air. Molecules can undergo these same kinds of motions, and do so when you supply them with energy. As collections of molecules undergo changes in energy, those collections move through the states of matter — solid, liquid, and gas. The body of ideas that explains all this is called *kinetic theory*.

Kinetic theory first made a name for itself when scientists attempted to explain and predict the properties of gases, and, in particular, how those properties changed with varying temperature and pressure. The idea emerged that the particles of matter within a gas (atoms or molecules) undergo a serious amount of motion as a result of the kinetic energy within them.

Kinetic energy is the energy of motion. Gas particles have a lot of kinetic energy, and constantly zip about, colliding with one another or with other objects. This is a complicated picture, but scientists simplified things by assuming that all the motions of the particles were *random*, that the only motion was *translation* (moving from place to place, as opposed to twisting, vibrating, and spinning), and that when particles collided, the collisions were *elastic* (perfectly bouncy, with no loss of energy). The gas particles were assumed to *neither attract nor repel* one another. Gases that actually behave in this simplified way are called *ideal*. The model of ideal gases explains why gas pressure increases with increased temperature. By heating a gas, you add kinetic energy to the particles; as a result, the particles collide with greater force upon other objects, so those objects experience greater pressure. (Check out Chapter 11 for full details on gas laws.)

When atoms or molecules have less kinetic energy, or when that energy must compete with other effects (like high pressure or strong attractive forces), the matter ceases to be in the diffuse, gaseous state and comes together into one of the condensed states: liquid or solid. Here are the differences between the two:

- ✔ The particles within a liquid are much closer together than those in a gas. As a result, applying pressure to a liquid does very little to change the volume. The particles still have an appreciable amount of kinetic energy associated with them, so they may undergo various kinds of twisting, stretching, and vibrating motions. In addition, the particles can slide past one another (translate) fairly easily, so liquids are fluid, though less fluid than gases. Fluid matter assumes the shape of anything that contains it.

- ✔ The state of matter with the least amount of kinetic energy is the solid. In a solid, the particles are packed together quite tightly and undergo almost no translation. Therefore, solids are not fluid. Matter in the solid state may still vibrate in place or undergo other low-energy types of motion, however, depending on its temperature (in other words, on its kinetic energy).

The temperatures and pressures at which different types of matter switch between states depend on the unique properties of the atoms or molecules within that matter. Typically, particles that are very attracted to one another and have easily stackable shapes tend toward condensed states. Particles with no mutual attraction (or that have mutual repulsion) and with not so easily stackable shapes tend toward the gaseous state. Think of a football game between fiercely rival schools. When fans of either school sit in their own section of the stands, the crowd is orderly, sitting nicely in rows. Put rival fans into the same section of the stands, however, and they'll repel each other with great energy.

Q. Are real-life gases more likely to behave like ideal gases at very high pressure or very low pressure, and why?

A. Real-life gases are more likely to behave like ideal gases when they are at very low pressure. Under very high pressure conditions, gas particles are much closer to one another and undergo greater numbers of collisions. Because of their close proximity, real-life gas particles are more likely to experience mutual attractions or repulsions. Because of their more frequent collisions, gases at very high pressure are more likely to reveal the effects of any inelasticity in their collisions (losing energy upon colliding).

1. Why does it make sense that the noble gases (which we introduce in Chapter 4) exist as gases at normal temperature and pressure?

Solve It

2. Ice floats in water. Based on the usual assumptions of kinetic theory, why is this weird?

Solve It

3. At the same temperature, how would the pressure of an ideal gas differ from that of a gas with mutually attractive particles? How would it differ from that of a gas with mutually repulsive particles?

Solve It

Make a Move: Figuring Out Phase Transitions and Diagrams

Each state (solid, liquid, gas) is called a *phase*. When matter moves from one phase to another because of changes in temperature and/or pressure, that matter is said to undergo a *phase transition*. Moving from liquid to gas is called *boiling,* and the temperature at which boiling occurs is called the *boiling point.* Moving from solid to liquid is called *melting,* and the temperature at which melting occurs is called the *melting point.* The melting point is the same as the *freezing point,* but freezing implies matter moving from liquid to solid phase.

At the surface of a liquid, molecules can enter the gas phase more easily than elsewhere within the liquid because the motions of those molecules aren't as constrained by the molecules around them. So, these surface molecules can enter the gas phase at temperatures below the liquid's characteristic boiling point. This low-temperature phase change is called *evaporation* and is very sensitive to pressure. Low pressures allow for greater evaporation, while high pressures encourage molecules to re-enter the liquid phase in a process called *condensation.* The pressure of the gas over the surface of a liquid is called the *vapor pressure.* Understandably, liquids with low boiling points tend to have high vapor pressures because the particles are weakly attracted to each other. At the surface of a liquid, weakly interacting particles have a better chance to escape into the vapor phase, thereby increasing the vapor pressure. See how kinetic theory helps make sense of things?

At the right combination of pressure and temperature, matter can move directly from solid to a gas or vapor. This type of phase change is called *sublimation,* and is the kind of phase change responsible for the white mist that emanates from dry ice, the common name for solid carbon dioxide. Movement in the opposite direction, from gas directly into solid phase, is called *desublimation.*

For any given type of matter there is a unique combination of pressure and temperature at the nexus of all three states. This pressure-temperature combination is called the *triple point.* At the triple point, all three phases coexist. In the case of good old H_2O, going to the triple point would produce boiling ice water. Take a moment to bask in the weirdness.

Other weird phases include plasma and supercritical fluids.

✔ *Plasma* is a gas-like state in which electrons pop off gaseous atoms to produce a mixture of free electrons and cations (atoms or molecules with positive charge). For most types of matter, achieving the plasma state requires very high temperatures, very low pressures, or both. Matter at the surface of the sun, for example, exists as plasma.

✔ *Supercritical fluids* exist under high temperature–high pressure conditions. For a given type of matter, there is a unique combination of temperature and pressure called the *critical point.* At temperatures and pressures higher than those at this point, the phase boundary between liquid and gas disappears, and the matter exists as a kind of liquidy gas or gassy liquid. Supercritical fluids can diffuse through solids like gases do, but can also dissolve things like liquids do.

Phase diagrams are useful tools for describing the states of a given type of matter across different temperatures and pressures. A phase diagram usually displays changes in temperature on the horizontal axis, and changes in pressure on the vertical axis. Lines drawn within the temperature-pressure field of the diagram represent the boundaries between phases, as shown for water and carbon dioxide in Figure 10-1.

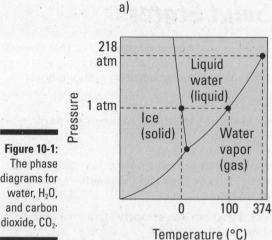

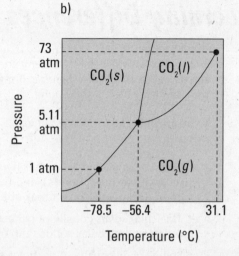

Figure 10-1:
The phase diagrams for water, H$_2$O, and carbon dioxide, CO$_2$.

Q. Ethanol (C$_2$H$_6$O) has a freezing point of –114°C. 1-Propanol (C$_3$H$_8$O) has a melting point of –88°C. At 25°C (where both compounds are liquids), which one is likely to have the higher vapor pressure, and why?

A. Ethanol has the higher vapor pressure at 25°C. First, notice that a freezing point and a melting point are the same thing — the temperature at which a substance undergoes the liquid-solid or solid-liquid phase transition. Next, compare the freezing/melting points of ethanol and propanol. Much colder temperatures must be achieved to freeze ethanol than to freeze propanol. This suggests that ethanol molecules have fewer attractive forces between themselves than do propanol molecules. At 25°C, both compounds are in liquid phase. Pure liquids in which the particles have less intermolecular (between-molecule) attraction have higher vapor pressure, because it's easier for molecules at the surface of the liquid to escape into vapor (gas) phase.

4. A cup of water is put into a freezer and cools to the solid phase within an hour. The water remains at that temperature for six months. After six months, the cup is retrieved from the freezer. The cup is empty. What happened?

Solve It

5. A sample of carbon dioxide is heated and pressurized within a container until it becomes a supercritical fluid. Then some carbon dioxide is allowed to escape the container while temperature is held constant. What is the most likely outcome?

Solve It

Discerning Differences in Solid States

Solids all have less kinetic energy than their liquid or gaseous counterparts, but that doesn't mean all solids are alike. Make any such claim and you'll instantly offend a whole class of scientists called *solid state chemists*. It may be that this is an overly sensitive group to begin with, but they have a solid point.

To begin with, the properties of a solid depend heavily on the forces between the particles within it.

- *Ionic solids* are held together by an array of very strong ionic bonds (see Chapter 5 for more about these bonds), and therefore tend to have high melting points — it takes a great deal of energy to break apart the particles.

- *Molecular solids* consist of packed molecules that are less strongly attractive to each other, so molecular solids tend to have lower melting points.

- Some solids consist of particles that are covalently bonded to one another; these *covalent solids* tend to be exceptionally strong because of the strength of their extensive covalent bond network; one example of a covalent solid is diamond. Covalent solids have *very* high melting points. Ever try to melt a diamond? (Chapter 5 has details about covalent bonds.)

In addition, the degree of order in the packing of particles within a solid can vary tremendously.

- Most solids are highly ordered, packing into neat, repeating patterns called *crystals*. The smallest packing unit, the one that repeats over and over to form the *crystalline solid,* is called the *unit cell.* Crystalline solids tend to have well-defined melting points.

- *Amorphous solids* are those solids that lack an ordered packing structure. Glass and plastic are examples of amorphous solids. Amorphous solids tend to melt over a broad range of temperatures.

When cooling a liquid through a phase transition into a solid, the rate of cooling can have a significant impact on the properties of the solid. The extreme order with which particles are packed together in crystalline solids can take lots of time. So, substances that are capable of forming crystalline solids may nevertheless freeze into amorphous solids if they're cooled rapidly. The particles may become trapped in disordered packing arrangements.

Q. At normal pressures, H_2 melts at $-259°C$, H_2O melts at $0°C$, NaCl melts at $801°C$, and diamond (pure carbon) melts at $3,550°C$. What are the different properties of these compounds that best explain such a wide range of melting points?

A. H_2 is a completely nonpolar molecule that forms a molecular solid; the interactions between H_2 molecules within this solid are very weak, so even small amounts of heat energy disrupt them. H_2O also forms a molecular solid, but H_2O is a highly polar molecule; within the H_2O solid, molecules align themselves to take maximum advantage of attractive forces between dipoles. More heat must be added to disrupt these interactions than is required to melt H_2. NaCl is an ionic solid, composed of a lattice of alternating Na^+ cations and Cl^- anions; the attractive forces between ions of opposite charge are strong, so it takes a lot of heat energy to disrupt them. Diamond is a covalent solid, composed of a network of covalently interconnected carbon atoms. Each carbon atom is covalently bound to three others; it takes a tremendous amount of heat energy to disrupt these bonds.

6. A liquid sample is divided into two batches. The first batch is very rapidly cooled to a temperature below the freezing point. The second batch is cooled to the same temperature, but very slowly. Then the two batches are slowly reheated. The melting point of one batch is measured to be precisely 280K. The other batch appears to melt gradually over 270–290K. What explains these observations?

7. Chemists can sometimes figure out the three-dimensional structure of molecules by making a pure crystalline sample of the molecule, shining an X-ray beam at the crystal, and making very precise measurements of the way the X-rays interact with the crystal. These kinds of experiments are often performed while the crystal is cooled under a stream of liquid nitrogen. What might be the point of this nitrogen cooling?

Solve It

Answers to Questions on Changes of State

By this point in the chapter, you may feel depleted of kinetic energy. But that's okay, because it means that you're more ordered, right? See whether that lovely order is reflected in your answers to the practice questions.

1 The noble gases are described as noble because, due to their full valence shells, they have very low reactivity. Only very weak forces occur between the atoms of noble gas. The particles don't significantly attract one another under any but the most extreme (high pressure, low temperature) conditions. Therefore, only a small amount of heat is required to push these elements into the gas phase.

2 Kinetic theory describes matter as moving from solid to liquid phase (melting) as you add energy to the sample. The added energy causes the particles to undergo greater motion, and to collide with other particles more energetically. Usually, this means that a liquid is less dense than a solid sample of the same material, because the greater motion of the liquid particles prevents close packing. Solid water (ice), on the other hand, is less dense than liquid water because of the unique geometry of water crystals. Because solid water is less dense than liquid water, ice floats in water.

3 **At a given temperature, an ideal gas would exert greater pressure than a gas with mutually attractive particles and lower pressure than a gas with mutually repulsive particles.** On average, mutual attraction allows particles to occupy a smaller volume at a given kinetic energy, while mutual repulsion causes particles to attempt to occupy a larger volume.

4 **The frozen water sublimed, moving directly from a solid to a gaseous state.** This process occurs slowly at the temperatures and pressures found within normal household freezers, but does occur. Try it.

5 **The sample would most likely move from supercritical fluid to gas.** Allowing carbon dioxide to escape (while holding the temperature constant) results in decreased pressure, corresponding to a downward vertical movement on the phase diagram of carbon dioxide.

6 Although the two batches consisted of the same type of matter, they froze into different types of solids because of the difference in cooling rate. The rapidly cooled sample froze into an amorphous solid because kinetic energy was removed from the particles more rapidly than they could order themselves into a crystalline state. The slowly cooled sample froze into a crystalline solid. The amorphous solid melted over a broad range of temperatures, whereas the crystalline solid melted at a well-defined meting point.

7 Cooling the crystals maintains the crystalline packing order by keeping the molecules in a state of very low kinetic energy. Shining an X-ray beam into the sample adds energy to the molecules, threatening to disrupt their packing order. Simultaneously cooling the crystal maintains the order. Because the order of the crystal is maintained, the collected X-ray data can be more easily interpreted to figure out the structure of the molecules within the crystal.

Chapter 11

Obeying Gas Laws

● ●

In This Chapter

▶ Boiling down the basics of vapor pressure

▶ Seeing how pressure, volume, temperature, and moles work and play together

▶ Diffusing and effusing at different rates

● ●

At first pass, gases may seem to be the most mysterious of the states of matter. Nebulous and wispy, gases easily slip through our grip. For all their diffuse fluidity, however, gases are actually the best understood of the states. The key thing to understand about gases is that they tend to behave in the same ways — physically, if not chemically. For example, gases expand to fill the entire volume of any container in which you put them. Also, gases are easily compressed into smaller volumes. Even more so than liquids, gases easily form homogenous mixtures. Because so much open space occurs between individual gas particles, these particles are pretty laid back about the idiosyncrasies of their neighbors.

Chapter 10 sets down the basic assumptions of the kinetic molecular theory of gases, a set of ideas that explains gas properties in terms of the motions of gas particles. In summary, kinetic molecular theory describes the properties of ideal gases, ones that conform to the following criteria:

✔ Ideal gas particles have a volume that is insignificant compared to the volume the gas occupies as a whole. The relatively small volume of a 20-ounce soda bottle, for example, completely dwarfs the individual gas particles inside the bottle, making their sizes irrelevant to any ideal gas calculation.

✔ Ideal gases consist of very large numbers of particles in constant random motion.

✔ Ideal gas particles are neither attracted to one another nor repelled by one another.

✔ Ideal gas particles exchange energy only by means of perfectly elastic collisions — collisions in which the total kinetic energy of the particles remains constant.

Like all ideals (the ideal job, the ideal mate, and so on), ideal gases are entirely fictional. All gas particles occupy some volume. All gas particles have some degree of inter-particle attraction or repulsion. No collision of gas particles is perfectly elastic. But lack of perfection is no reason to remain unemployed or lonely. Neither is it a reason to abandon kinetic molecular theory of ideal gases. In this chapter, you're introduced to a wide variety of applications of kinetic theory, which come in the form of the so-called "gas laws."

Getting the Vapors: Evaporation and Vapor Pressure

Because kinetic molecular theory explains the movement of particles between different phases, it comes as no surprise that we find gases in the presence of liquids. Within a liquid, individual particles have any of a range of kinetic energies. At any given moment, some fraction of particles on the surface of the liquid possesses enough kinetic energy to escape into the gas phase. Collections of these gas-phase "refugee" particles, hovering over the surface of a liquid, are called *vapor*. The process by which particles escape the surface of a liquid into the vapor phase is called *evaporation*. Of course, the escaped particles may fall back into captivity in a process called *condensation*. Given enough time, the rate at which particles escape comes to equal the rate at which they are recaptured, and this state of affairs is called a *dynamic equilibrium*.

Vapor pressure is the pressure at which vapor is in dynamic equilibrium with liquid. As evaporation occurs, the pressure above the surface of a liquid increases because more particles are added to that volume. Eventually, the pressure over the surface of the liquid equals the vapor pressure of that liquid, at which point evaporation equals condensation. Why does increased pressure produce condensation? What is pressure, anyway? In short, *pressure* is a force applied over an area — like the area of the surface of a liquid, for example. At the molecular scale, pressure arises from the collisions of countless particles with that surface. At higher pressure, then, more gas particles are colliding more frequently with the surface of the liquid. These collisions make it more difficult for liquid-phase particles to escape and make it more likely that gas-phase particles will be recaptured. The final important piece of information to retain about vapor pressure is its relationship to boiling; specifically, when the vapor pressure of a liquid reaches the atmospheric pressure, a liquid will boil.

Vapor pressure depends on temperature because temperature affects the dynamic equilibrium. Adding heat to a sample increases the average kinetic energy of the particles within that sample. Temperature is simply a measure of the average kinetic energy of the particles in a sample. So at higher temperatures, particles possess greater kinetic energy and are more likely to escape into vapor phase. The higher the temperature, the higher the vapor pressure.

Pressure is also directly related to altitude. If you've ever tried to take a deep breath from a mountaintop, you know that the air is "thinner" at higher elevation, making it harder to get sufficient oxygen. This is because the relative atmospheric pressure is lower than at sea level, and fewer oxygen particles are in a given volume of air.

Q. Legend has it that watched pots do not boil. Rigorous experimentation has proved this maxim to be false. Some kitchen theorists claim that placing a tight-fitting lid on a pot (to prevent unauthorized pot-watching) actually causes water to boil at a higher temperature than does water in a pot with no lid. Is this claim plausible? Why or why not?

A. The claim is plausible. Placing a tight-fitting lid on a pot of water traps water vapor inside. Because the lid traps vapor inside, the pressure in the space above the water increases beyond the atmospheric pressure in the kitchen. As heat is added to the water, the vapor pressure increases. Boiling occurs only when the vapor pressure of the water reaches and/or exceeds the pressure over the surface of the liquid. So, water in the lidded pot must be heated more than water in an open pot before it will boil.

Of course, water in a lidded pot may nevertheless boil *more quickly,* because the lid also traps heat that would otherwise escape.

1. In a little hut on top of Mount Fuji in Japan, Tomoko boils water for her breakfast. Meanwhile, atop Half Dome in Yosemite National Park, Jim boils water, too. If Tomoko's water boils at 86°C and Jim's water boils at 90°C, which of the two is at the higher altitude (and therefore the more hardcore hiker)?

Solve It

2. On the planet Blurgblar, pressure decreases linearly with altitude. A Blurgblarian yodeler prepares tea atop the 2,250m peak of Mount J-11. The yodeler notes that the water boils at 283K. Meanwhile, at sea level, a Blurgblarian chemist measures the boiling point of water to be 298K. An unfortunate Earthling physicist suddenly emerges from a time-space wormhole on the surface of Blurgblar. Gasping at the cyanide-laced atmosphere, the physicist observes that a nearby Blurgblarian Boy Scout is boiling water at 293K. At what altitude does the physicist find himself?

Solve It

Playing with Pressure and Volume: Boyle's Law

Although the explanation of vapor pressure and evaporation endures for only a few paragraphs in the previous section, it mentions four important variables: pressure, volume, temperature, and the number of particles. Relationships between these four factors are the domain of the gas laws. We take a look at these in this section and the rest of the chapter.

The first of these relationships to have been formulated into a law is that between pressure and volume. Robert Boyle, an Irish gentleman regarded by some as the first chemist (or "chymist," as his friends might have said), is typically given credit for noticing that gas pressure and volume have an inverse relationship:

Volume = Constant × (1/Pressure)

This statement is true when the other two factors, temperature and number of particles, are fixed. Another way to express the same idea is to say that although pressure and volume may change, they do so in such a way that their product remains constant. So, as a gas undergoes change in pressure (P) and volume (V) between two states, the following is true:

$$P_1 \times V_1 = P_2 \times V_2$$

The relationship makes good sense in light of kinetic molecular theory. At a given temperature and number of particles, more collisions will occur at smaller volumes. These increased collisions produce greater pressure. And vice versa. Boyle had some dubious ideas about alchemy, among other things, but he really struck gold with the pressure-volume relationship in gases.

Q. A sealed plastic bag is filled with 1L of air at standard temperature and pressure (STP). You accidentally sit on the bag. The maximum pressure the bag can withhold before popping is 500 kilopascals (kPa). What is the internal volume of the bag at the instant of popping?

A. **0.2L.** The problem tells us that the bag has an initial volume, V_1, of 1L, and an initial pressure, P_1, of 101.3 kPa (the pressure at STP). (See Chapter 7 if you've

momentarily forgotten the definition of STP.) The pressure inside the bag reaches 500 kPa before popping, so that value represents P_2. So, the only missing variable is the final volume. Solve for the final volume, V_2, by plugging in the known values:

$$V_2 = (1L \times 101.3 \text{ kPa})/500 \text{ kPa} = 0.2L$$

3. An amateur entomologist captures a particularly excellent ladybug specimen in a plastic jar. The internal volume of the jar is 0.5L, and the air within the jar is initially at 1 atm. The bug-lover is so excited by the catch that he squeezes the jar fervently in his sweaty palm, compressing it such that the final pressure within the jar is 1.25 atm. What is the final volume of the ladybug's prison?

Solve It

4. A container possesses 3L internal volume. This volume is divided equally in two by a gas-tight seal. On one half of the seal, neon gas resides at 5 atm. The other half of the container is kept under vacuum. Suddenly and with great fanfare, the internal seal is broken. What is the final pressure within the container?

Solve It

Tinkering with Volume and Temperature: Charles's Law and Absolute Zero

Lest the Irish have all the gassy fun, the French contributed a gas law of their own. History attributes this law to French chemist Jacques Charles. Charles discovered a direct, linear relationship between the volume and the temperature of a gas:

Volume = Constant × Temperature

This statement is true when the other two factors, pressure and number of particles, are fixed. Another way to express the same idea is to say that although temperature and volume may change, they do so in such a way that their ratio remains constant. So, as a gas undergoes change in temperature (T) and volume (V) between two states, the following is true:

$V_1 / T_1 = V_2 / T_2$

Not to be outdone by the French, another Irish scientist took Charles's observations and ran with them. William Thomson, eventually to be known as Lord Kelvin, took stock of all the data available in his mid-19th century heyday and noticed a few things:

✔ First, plotting the volume of a gas versus its temperature always produced a straight line.

✔ Second, extending these various lines caused them all to converge at a single point, corresponding to a single temperature at zero volume. This temperature — though not directly accessible in experiments — was about –273 degrees Celsius. Kelvin took the opportunity to enshrine himself in the annals of scientific history by declaring that temperature as *absolute zero,* the lowest temperature possible.

This declaration had at least two immediate benefits. First, it happened to be correct. Second, it allowed Kelvin to create the Kelvin temperature scale, with absolute zero as the Official Zero. Using the Kelvin scale (where °C = K – 273), everything makes a whole lot more sense. For example, doubling the Kelvin temperature of a gas doubles the volume of that gas. When you work with Charles's Law, converting Celsius temperatures to Kelvin is crucial.

Q. A red rubber dodge ball sits in a 20.0°C basement, filled with 3.50L of compressed air. Eager to begin practice for the impending dodge ball season, Vince reclaims the ball and takes it outside. After a few hours of practice, the well-sealed ball has a volume of 3.00L. What's the temperature outside?

A. **–22°C.** The question provides an initial temperature, an initial volume, and a final volume. You're asked to find the final temperature, T_2. Apply Charles's Law, plugging in the known values and solving

for the final temperature. But take care — Charles's Law requires you to convert all temperatures to K (where K = °C + 273). After a few hours outdoors, it's safe to assume that the temperature of the ball has reached equilibrium. Because the volume of the ball decreased, expect the outdoor temperature to be lower than the temperature in the basement:

$T_2 = (293K \times 3.00L) / 3.50L = 251K$

This temperature corresponds to –22°C. Vince may have some judgment issues.

5. Jacques Charles's ghost attempts to inflate his sagging celestial hot air balloon with short bursts from a burner. The initial volume of the balloon is 300L (ghosts don't require large balloons). Trying to impress the passing spirit of a recently departed *séductrice,* Jacques lets fly a long burst from the burner. The long burn increases the temperature of the air within in the balloon from 40.0°C to 50.0°C. How much does the balloon inflate?

Solve It

6. Always helpful, Danny persuades his little sister Suzie that her Very Special Birthday Balloon will last much longer if she puts it in the freezer in the basement for a while. The temperature in the house is 20.0°C. The balloon has an initial volume of 0.250L. If the balloon has collapsed to 0.200L by the time Suzie catches on to Danny's devious deed, what is the temperature inside the freezer?

Solve It

All Together Now: The Combined and Ideal Gas Laws

Boyle's and Charles's laws are convenient if you happen to find yourself in situations where only two factors change at a time. The universe is rarely so well-behaved. What if pressure, temperature, and volume all change at the same time? Is aspirin and a nap the only solution? No. Enter the *Combined Gas Law:*

$$\frac{P_1 \times V_1}{T_1} = \frac{P_2 \times V_2}{T_2}$$

Of course, the real universe can fight back by changing another variable. In the real universe, for example, tires spring leaks. In such a situation, gas particles escape the confines of the tire. This escape decreases the number of particles, *n,* within the tire. Cranky, tire-iron wielding motorists on the side of the road will attest that decreasing *n* decreases volume. This relationship is sometimes expressed as *Avogadro's Law:*

Volume = Constant × Number of particles

Combining Avogadro's Law with the Combined Gas Law produces the wonderfully comprehensive relationship:

$$\frac{P_1 \times V_1}{T_1 \times n_1} = \frac{P_2 \times V_2}{T_2 \times n_2}$$

The final word on ideal gas behavior summarizes all four variables (pressure, temperature, volume, and number of particles) in one easy-to-use equation called the *Ideal Gas Law:*

$$PV = nRT$$

Here, R is the gas constant, the one quantity of the equation that can't change. Of course, the exact identity of this constant depends on the units you're using for pressure, temperature, and volume. A very common form of the gas constant as used by chemists is $R = 0.08206$L atm K^{-1} mol^{-1}. Alternately, you may encounter $R = 8.314$L kPa K^{-1} mol^{-1}.

Q. A 0.80L container holds 10 mol of helium. The temperature of the container is 10°C. What's the internal pressure of the container?

A. 2.9×10^2 **atm.** Consider your known and unknown variables. You're given volume, number of particles, and temperature. You're asked to calculate pressure. The equation that fills the bill is the Ideal Gas Law, $PV = nRT$. Rearrange the equation to solve for P, so that $P = (nRT) / V$. Now, before you blithely plug your known values into the equation, be sure that all

your units agree with those used in the gas constant you've chosen. Here, we use $R = 0.08206$L atm K^{-1} mol^{-1}. So, we must convert the temperature (10°C) into Kelvin, K = 10 + 273 = 283 K. Next, plug in your known values and solve:

Pressure = (10 mol × 0.08206L atm K^{-1} mol^{-1} × 283K) / 0.80L = 2.9×10^2 atm

That's nearly 300 times normal atmospheric pressure. Stay away from that container.

---

7. The 0.80L container from the example question breaks a seal. Because the container stored a poisonous gas, it was itself stored within a larger, vacuum-sealed container. After the poisonous gas expands to fill the newly available volume, the gas is at STP. What is the total volume of the secondary container?

Solve It

8. A container with a volume of 15.0L contains oxygen. The gas is at a temperature of 29.0°C and a pressure of 1.00×10^4 kPa. How many moles of gas occupy that container?

Solve It

9. The volume of a whoopee cushion is 0.450L at 27.0°C and 105 kPa. Danny has placed one such practical joke device on the chair of his unsuspecting Aunt Bertha. Unbeknownst to Danny, this particular whoopee cushion suffers from a construction defect that sometimes blocks normal out-gassing and ruins the flatulence effect. So, even when the cushion receives the full force of Aunt Bertha's ample behind, the blockage prevents deflation. The cushion sustains the pressure exerted by Bertha, so that the internal pressure becomes 200 kPa. As she sits on the cushion, Bertha warms its contents a full 10.0°C. At last, and to Danny's profound satisfaction, the cushion explodes. What volume of air does it expel?

Solve It

Mixing It Up with Dalton's Law of Partial Pressures

REMEMBER

Gases mix. They do so better than liquids and infinitely better than solids. So, what's the relationship between the total pressure of a gaseous mixture and the pressure contributions of the individual gases? Here is a satisfyingly simple answer: Each individual gas within the mixture contributes a partial pressure, and adding the partial pressures yields the total pressure. This relationship is summarized by *Dalton's Law of Partial Pressures,* for a mixture of individual gases:

$$P_{total} = P_1 + P_2 + P_3 + \ldots + P_n$$

This relationship makes sense if you think about pressure in terms of kinetic molecular theory. Adding a gaseous sample into a volume that already contains other gases increases the number of particles in that volume. Because pressure depends on the number of particles colliding with the container walls, increasing the number of particles increases the pressure proportionally.

EXAMPLE

Q. A chemist designs an experiment to study the chemistry of the atmosphere on the early earth, billions of years ago. She constructs an apparatus to combine pure samples of the primary volcanic gases that made up the early atmosphere: carbon dioxide, ammonia, and water vapor. If the partial pressures of these gases are 50 kPa, 80 kPa, and 120 kPa, respectively, what's the pressure of the resulting mixture?

A. 250 kPa. However difficult early earth atmospheric chemistry may prove to be, this particular problem is a simple one. Dalton's Law states that the total pressure is simply the sum of the partial pressures of the component gases:

$$\begin{aligned} P_{total} &= P(CO_2) + P(NH_3) + P(H_2O) \\ &= 50 \text{ kPa} + 80 \text{ kPa} + 120 \text{ kPa} \\ &= 250 \text{ kPa} \end{aligned}$$

10. A chemist adds solid zinc powder to a solution of hydrochloric acid to initiate the following reaction:

$$Zn(s) + 2\ HCl(aq) \rightarrow ZnCl_2(aq) + H_2(g)$$

The chemist inverts a test tube and immerses the open mouth into the reaction beaker to collect the hydrogen gas that bubbles up from the solution. The reaction proceeds to equilibrium. At the end of the experiment, the water levels within the tube and outside the tube are equal. The pressure in the lab is 101.3 kPa, and the temperature of all components is 298K. The vapor pressure of water at 298K is 3.17 kPa. What is the partial pressure of dihydrogen gas trapped in the tube?

Solve It

Diffusing and Effusing with Graham's Law

"Wake up and smell the coffee." This command is usually issued in a scornful tone, but most people who have awakened to the smell of coffee remember the event fondly. The morning gift of coffee aroma is made possible by a phenomenon called *diffusion.* Diffusion is the movement of a substance from an area of higher concentration to an area of lower concentration. Diffusion occurs spontaneously, on its own. Diffusion leads to mixing, eventually producing a homogenous mixture in which the concentration of any gaseous component is equal throughout an entire volume. Of course, that state of complete diffusion is an equilibrium state; achieving equilibrium can take time.

Different gases diffuse at different rates, depending on their molar masses (see Chapter 7 for details on molar masses). The rates at which two gases diffuse can be compared using *Graham's Law.* Graham's Law also applies to *effusion,* the process in which gas molecules flow through a small hole in a container. Whether gases diffuse or effuse, they do so at a rate inversely proportional to the square root of their molar mass. In other words, more massive gas molecules diffuse and effuse more slowly than less massive gas molecules. So, for gases A and B:

$$\frac{Rate_A}{Rate_B} = \frac{\sqrt{molar\ mass_B}}{\sqrt{molar\ mass_A}}$$

Q. How much faster does hydrogen gas effuse than neon gas?

A. **3.2 times faster.** First, "hydrogen gas" refers to dihydrogen, H_2. Next, consult your periodic table (or your memory, if you're that good) to obtain the molar masses of dihydrogen gas ($2.0g \ mol^{-1}$) and neon gas ($20g \ mol^{-1}$). Finally plug those values into the appropriate places within Graham's Law.

$$\frac{\text{Rate } H_2}{\text{Rate Ne}} = \frac{\sqrt{20}}{\sqrt{2.0}} = 3.2$$

So, dihydrogen effuses 3.2 times faster than neon.

---

11. Mystery Gas A effuses 4.0 times faster than oxygen. What is it the likely identity of the Mystery Gas?

Solve It

Answers to Questions on Gas Laws

You've answered the practice questions on gas behavior. Were your answers ideal? Check them here. No pressure.

1. **Tomoko is at the higher altitude.** As altitude increases, atmospheric pressure decreases. As temperature increases, vapor pressure increases. Liquids boil when their vapor pressure exceeds the external (atmospheric) pressure. So, because Tomoko's water boils at a lower temperature, she must be at a higher altitude, where atmospheric pressure is lower. Tomoko is the hardcore hiker.

2. **750m.** Boiling point decreases linearly with external pressure. Because the atmospheric pressure on Blurgblar decreases linearly with altitude, boiling point and altitude also have a linear relationship on this odd planet. To calculate the altitude where the physicist emerged, you must determine the linear relationship between altitude and boiling point. The total elevation increase between sea level and the peak of Mount J-11 is 2,250m. The decrease in boiling point over that change in altitude is –15K (298K – 283K). So, the boiling point decreases by 1K for every 150m increase in altitude:

   2,250m / 15K = 150 m/K

   The dying physicist observed a boiling point 5K lower than the one observed at sea level. So, the physicist is 850m above sea level:

   5K × 150 m/K = 750m

3. **0.4L.** You're given an initial pressure, an initial volume, and a final pressure. Boyle's Law leaves you with one unknown: final volume. Solve for the final volume by plugging in the known values:

   V_2 = (1 atm × 0.5L) / 1.25 atm = 0.4L

4. **2.5 atm.** Under the initial conditions, gas at 5 atm resides in a 1.5L volume. When the seal is removed, the entire 3L of the container becomes available to the gas, which expands to occupy the new volume. Predictably, its pressure decreases. To calculate the new pressure, P_2, plug in the known values and solve:

   P_2 = (5 atm × 1.5L) / 3L = 2.5 atm

5. **10L.** You're given the initial volume and initial temperature of the balloon, as well as the balloon's final temperature. Apply Charles's Law, plugging in the known values and solving for the final volume. Be careful — all temperatures must be expressed in units of K:

   V_2 = (300L × 323K) / 313K = 310L

   The difference in volume due to heat-induced inflation is 310L – 300L = 10L.

6. **–39°C.** Charles's Law is the method here. The unknown is the final temperature, T_2. You're given an initial temperature, as well as the initial and final volumes. After converting the temperatures to units of K, plug in the known values and solve for final temperature:

   T_2 = (293K × 0.200L) / 0.250L = 234K

   This final temperature corresponds to –39°C. That's one serious freezer.

7. **220L.** The number of moles of gas (10 mol) remains constant. The other three factors (pressure, temperature, and volume) all change between initial and final states. So, you need to use the Combined Gas Law. The initial values ($2.9 × 10^2$ atm, 283K, 0.80L) all derive from the example problem. The final temperature and pressure are known (273K, 1 atm) because the question states that the gas ends up at STP. So, the only unknown is the final volume. Rearrange the

Combined Gas Law to solve for this value:

$$V_2 = \frac{P_1 \times V_1 \times T_2}{T_1 \times P_2} = \frac{2.9 \times 10^2 \text{ atm} \times 0.80\text{L} \times 273\text{K}}{283\text{K} \times 1 \text{ atm}} = 2.2 \times 10^2 \text{ atm}$$

8 **59.7mol.** This problem simply requires use of the Ideal Gas Law, arranged to solve for number of moles, n. Don't forget to convert temperature to units of K and to use the appropriate version of the gas constant, R.

$$n = \frac{P \times V}{R \times T} = \frac{1.0 \times 10^4 \text{ kPa} \times 15.0\text{L}}{8.314 \dfrac{\text{kPa} \times \text{L}}{\text{mol} \times \text{K}} \times 302\text{K}} = 59.7 \text{ mol}$$

9 **0.244L.** You're given an initial volume, initial temperature, and initial pressure. You're also given a final pressure and a final temperature. The only unknown is final volume. Rearrange the Combined Gas Law to solve for final volume, V_2.

$$V_2 = \frac{P_1 \times V_1 \times T_2}{T_1 \times P_2} = \frac{105 \text{ kPa} \times 0.450\text{L} \times 310\text{K}}{300\text{K} \times 200 \text{ kPa}} = 0.244\text{L}$$

Despite the fact that the final temperature is higher than the initial temperature, the final volume is much smaller than the initial volume. In effect, 10 trifling degrees are no match for the pressure exerted by Bertha's posterior.

10 **98.1 kPa.** The system has come to equilibrium, so the interior of the tube contains a gaseous mixture of dihydrogen and water vapor. Because the water levels inside and outside the tube are equal, you know that the total pressure inside the tube equals the ambient pressure of the lab, 101.3 kPa. The total pressure includes the partial pressure contributions from dihydrogen and from water vapor. Set up an equation using Dalton's Law:

$P_{total} = P(H_2) + P(H_2O)$

Rearrange the equation to solve for $P(H_2)$ and substitute in the known values to solve:

$P(H_2) = 101.3 \text{ kPa} - 3.17 \text{ kPa} = 98.1 \text{ kPa}$

11 **Dihydrogen, H_2.** The question states that the ratio of the rates is 4.0. Recall that oxygen gas is dioxygen, O_2, with a molar mass of 32g mol^{-1}. Substitute these known values into Graham's Law.

$$\frac{\text{Rate A}}{\text{Rate O}_2} = 4.0 = \frac{\sqrt{32}}{\sqrt{\text{molar mass A}}}$$

Square both sides of this equation to bring values out from underneath the radicals.

$$16 = \frac{32}{\text{molar mass A}}$$

Next, rearrange to solve for the molar mass of Gas A:

Molar mass A = $(32/16)$g mol^{-1} = 2.0g mol^{-1}

This molar mass is consistent with dihydrogen, H_2.

Chapter 12

Dissolving into Solutions

Compounds can form mixtures. When compounds mix completely, right down to the level of individual molecules, we call the mixture a *solution*. Each type of compound in a solution is called a *component*. The component of which there is the most is usually called the *solvent*. The other components are called *solutes*. Although most people think "liquid" when they think of solutions, a solution can be a solid, liquid, or gas. The only criterion is that the components are completely intermixed. We explain what you need to know in this chapter.

Seeing Different Forces at Work in Solubility

For gases, forming a solution is a straightforward process. Gases simply diffuse into a common volume (see Chapter 11 for more about diffusion). Things are a bit more complicated for condensed states like liquids and solids. In liquids and solids, molecules or ions are crammed so closely together that *intermolecular forces* are very important. Examples of these kinds of forces include ion-dipole, dipole-dipole, hydrogen bonding, and London (dispersion) forces. We touch on the physical underpinnings of these forces in Chapter 5.

Introducing a solute into a solvent initiates a tournament of forces. Attractive forces between solute and solvent compete with attractive solute-solute and solvent-solvent forces. A solution forms only to the extent that solute-solvent forces dominate over the others. The process in which solvent molecules compete and win in the tournament of forces is called *solvation* or, in the specific case where water is the solvent, *hydration*. Solvated solutes are surrounded by solvent molecules. When solute ions or molecules become separated from one another and surrounded in this way, we say they're *dissolved*.

Imagine that the members of a ridiculously popular "boy band" exit their hotel to be greeted by an assembled throng of fans and the media. The band members attempt to cling to each other, but are soon overwhelmed by the crowd's ceaseless, repeated attempts to get closer. Soon, each member of the band is surrounded by his own attending shell of reporters and hyperventilating teenage girls. So it is with dissolution.

The tournament of forces plays out differently among different combinations of components. In mixtures where solute and solvent are strongly attracted to one another, more solute can be dissolved. One factor that always tends to favor dissolution is *entropy,* a kind of disorder or "randomness" within a system. Dissolved solutes are less ordered than undissolved solutes. Beyond a certain point, however, adding more solute to a solution doesn't result in a greater amount of solvation. At this point, the solution is in dynamic equilibrium; the rate at which solute becomes solvated equals the rate at which dissolved solute *crystallizes,* or falls out of solution. A solution in this state is *saturated.* By contrast, an *unsaturated* solution is one that can accommodate more solute. A *supersaturated* solution is one in which more solute is dissolved than is necessary to make a saturated solution. A supersaturated solution is unstable; solute molecules may crash out of solution given the slightest perturbation. The situation is like that of Wile E. Coyote who runs off a cliff and remains suspended in the air until he looks down — at which point he inevitably falls.

The concentration of solute required to make a saturated solution is the *solubility* of that solute. Solubility varies with the conditions of the solution. The same solute may have different solubility in different solvents, and at different temperatures, and so on.

When one liquid is added to another, the extent to which they intermix is called *miscibility.* Typically, liquids that have similar properties mix well — they are *miscible.* Liquids with dissimilar properties often don't mix well — they are *immiscible.* This pattern is summarized by the phrase, "like dissolves like." Alternately, you may understand miscibility in terms of the Italian Salad Dressing Principle. Inspect a bottle of Italian salad dressing that has been sitting in your refrigerator. Observe the following: The dressing consists of two distinct layers, an oily layer and a watery layer. Before using, you must shake the bottle to temporarily mix the layers. Eventually, they'll separate again because water is polar and oil is nonpolar. (See Chapter 5 if the distinction between polar and nonpolar is lost on you.) Polar and nonpolar liquids mix poorly, though occasionally with positive gastronomic consequences.

Similarity or difference in polarity between components is often a good predictor of solubility, regardless of whether those components are liquid, solid, or gas. Why is polarity such a good predictor? Because polarity is central to the tournament of forces that underlies solubility. So, solids held together by ionic bonds (the most polar type of bond) or polar covalent bonds tend to dissolve well in polar solvents, like water. For a refresher on ionic and covalent bonding, visit Chapter 5.

Q. Sodium chloride dissolves more than 25 times better in water than in methanol. Explain this difference, referring to the structure and properties of water, methanol, and sodium chloride.

A. Sodium chloride (NaCl) is an ionic solid, a lattice composed of sodium cations (atoms with positive charge) alternating with chlorine anions (atoms with negative charge). The lattice has a highly regular, idealized geometry and is held together by ionic bonds, the most polar type of bond. To dissolve NaCl, a solvent must be able to engage in very polar interactions with these ions and do so with near-ideal geometry. The structure and properties of water (which is polar) are better suited to this task than are those of methanol (see the following figure). The two O–H bonds of water (on the left) partially sum to produce a strong dipole along the mirror image plane of the molecule that runs between the two hydrogen atoms. Methanol (on the right) is also polar, due largely to its own O–H bond, but is less polar than water. In solution, water molecules can orient their dipoles cleanly and in either of two directions to interact favorably with Na⁺ or Cl⁻ ions. Methanol molecules can engage in favorable interactions with these ions too, but not nearly as well as water.

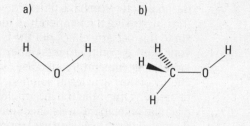

a)   b)

1. *Lattice energy* is a measure of the strength of the interactions between ions in the lattice of an ionic solid. The larger the lattice energy, the stronger the ion-ion interactions. Here is a table of ionic solids and their associated lattice energies. Predict the rank order of solubility in water of these ionic solids.

Sodium Salt	Lattice Energy, kJ mol^{-1}
NaBr	747
NaCl	787
NaF	923
NaI	704

Solve It

2. Ethanol, CH_3CH_2OH, is miscible with water. Octanol, $CH_3(CH_2)_7OH$, isn't miscible in water. Is sucrose (as in table sugar) likely to be more soluble in ethanol or octanol? Why?

Solve It

Altering Solubility with Temperature

REMEMBER

Increasing temperature magnifies the effects of entropy on a system. Because the entropy of a solute is usually increased when it dissolves, increasing temperature usually increases solubility — for solid and liquid solutes, anyway. Another way to understand the effect of temperature on solubility is to think about heat as a reactant in the dissolution reaction:

Solid solute + Water + Heat → Dissolved solute

Heat is usually absorbed when a solute dissolves. Increasing temperature corresponds to added heat. So, by increasing temperature you supply a needed reactant in the dissolution reaction. (In those rare cases where dissolution releases heat, increasing temperature can decrease solubility.)

Gaseous solutes behave differently than do solid or liquid solutes with respect to temperature. Increasing the temperature tends to decrease the solubility of gas in liquid. To understand this pattern, check out the concept of vapor pressure. (If creaking sounds emanate from your skull as you try to remember what vapor pressure is about, take a peek at Chapter 11.) Increasing temperature increases vapor pressure because added heat increases the kinetic energy of the particles in solution. With added energy, these particles stand a greater chance of breaking free from the intermolecular forces that hold them in solution. A classic, real-life example of temperature's effect on gas solubility is carbonated soda. Which goes flat (loses its dissolved carbon dioxide gas) more quickly: warm soda or cold soda?

The comparison of gas solubility in liquids with the concept of vapor pressure highlights another important pattern: Increasing pressure increases the solubility of a gas in liquid. Just as high pressures make it more difficult for surface-dwelling liquid molecules to escape into vapor phase, high pressures inhibit the escape of gases dissolved in solvent. The relationship between pressure and gas solubility is summarized by *Henry's Law:*

Solubility = Constant × Pressure

The "constant" is *Henry's Constant,* and its value depends on the gas, solvent, and temperature. A particularly useful form of Henry's Law relates the change in solubility (*S*) that accompanies a change in pressure (*P*) between two different states:

$$S_1 / P_1 = S_2 / P_2$$

According to this relationship, tripling the pressure triples the gas solubility, for example.

Q. Henry's Constant for dinitrogen gas in water at 293K is 0.69×10^{-3} mol L^{-1} atm^{-1}. The partial pressure of dinitrogen in air at sea level is 0.78 atm. What is the solubility of N_2 in a glass of water at 20°C sitting on a coffee table within a beach house?

A. **0.54×10^{-3} mol L^{-1}.** This problem requires the direct application of Henry's Law.

The glass of water is at 20°C, which is equivalent to 293K (just add 273 to any Celsius temperature to get the Kelvin equivalent). Because the glass sits within a beach house, we can assume the glass is at sea level. So, we can use the provided values for Henry's Constant and the partial pressure of N_2.

Solubility = $(0.69 \times 10^{-3}$ mol L^{-1} atm^{-1}) × (0.78 atm) = 0.54×10^{-3} mol L^{-1}

3. A chemist prepares an aqueous solution of cesium sulfate, $Ce_2(SO_4)_3$, swirling the beaker in her gloved hand to promote dissolution. She notices something, momentarily furrows her brow, and then smiles knowingly. She nestles the beaker into a bed of crushed ice within a bucket. What did the chemist notice, why was she briefly confused, and why did she place the dissolving cesium sulfate on ice?

4. Deep-sea divers routinely operate under pressures of multiple atm. One malady these divers must be concerned with is "the bends," a dangerous condition that occurs when divers rise too quickly from the depths, resulting in the over-rapid release of gas from blood and tissues. Why do the bends occur?

Solve It

5. Reefus readies himself for a highly productive Sunday afternoon of football watching, arranging bags of cheesy poofs and a six-pack of grape soda around his beanbag chair. At kickoff, Reefus cracks open his first grape soda and settles in for the long haul. Three hours later, covered in cheesy crumbs, Reefus marks the end of the fourth quarter by cracking open the last of the six-pack. The soda fizzes violently all over Reefus and the beanbag chair. What happened?

6. The grape soda preferred by Reefus (the gentleman introduced in Question 5) is bottled under 3.5 atm of pressure. Reefus lives on a bayou at sea level (hint: 1 atm). The temperature at which the soda is bottled is the same as the temperature in Reefus's living room. Assuming that the concentration of carbon dioxide in an unopened grape soda is $0.15 \ mol \ L^{-1}$, what is the concentration of carbon dioxide in an opened soda that has gone flat while Reefus naps after the game?

Solve It

Concentrating on Molarity and Percent Solutions

It seems that different solutes dissolve to different extents in different solvents in different conditions. How can anybody keep track of all these differences? Chemists do so by measuring *concentration*. Qualitatively, a solution with a large amount of solute is said to be *concentrated*. A solution with only a small amount of solute is said to be *dilute*. As you may suspect, simply describing a solution as concentrated or dilute is usually about as useful as calling it "pretty" or naming it "Fifi." We need numbers. Two important ways to measure concentration are *molarity* and *percent solution*.

Molarity relates the amount of solute to the volume of the solution:

Molarity = (moles of solute) / (liters of solution)

To calculate molarity, you may have to use conversion factors to move between units. For example, if you're given the mass of a solute in grams, use the molar mass of that solute to convert the given mass into moles. If you're given the volume of solution in cm^3 or some other unit, you need to convert that volume into liters.

The units of molarity are always $mol\ L^{-1}$. These units are often abbreviated as M and referred to as "molar." Thus, $0.25M$ KOH(aq) is described as "Point two-five molar potassium hydroxide" and contains 0.25 moles of KOH per liter of solution. Note that this does *not* mean that there are 0.25 moles KOH per liter of *solvent* (water, in this case) — only the final volume of the solution (solute plus solvent) is important in molarity.

Like other units, the unit of molarity can be modified by standard prefixes, as in millimolar (mM, $10^{-3}\ mol\ L^{-1}$) and micromolar (μM, $10^{-6}\ mol\ L^{-1}$).

Percent solution is another common way to express concentration. The precise units of percent solution typically depend on the phase of each component. For solids dissolved in liquids, mass percent is usually used:

$$Mass\ \% = 100\% \times \frac{mass\ of\ solute}{total\ mass\ of\ solution}$$

This kind of measurement is sometimes called a mass-mass percent solution because one mass is divided by another. Very dilute concentrations (as in the concentration of a contaminant in drinking water) are sometimes expressed as a special mass percent called *parts per million (ppm)* or *parts per billion (ppb)*. In these metrics, the mass of the solute is divided by the total mass of the solution, and the resulting fraction is multiplied by 10^6 (ppm) or by 10^9 (ppb).

Sometimes, the term *percent solution* is used to describe concentration in terms of the final volume of solution, instead of the final mass. For example:

- "5% $Mg(OH)_2$" can mean 5g magnesium hydroxide in 100 mL final volume. This is a mass-volume percent solution.

- "2% H_2O_2" can mean 2 mL hydrogen peroxide in 100 mL final volume. This is a volume-volume percent solution.

TIP

Clearly, it's important to pay attention to units when working with concentration. Only by observing which units are attached to a measurement can you determine whether you are working with molarity, mass percent, or with mass-mass, mass-volume, or volume-volume percent solution.

EXAMPLE

Q. Calculate the molarity and the mass-volume percent solution obtained by dissolving 102.9g H_3PO_4 into 642 mL final volume of solution. Be sure to use proper units. (Hint: 642 mL = 0.642L)

A. First, calculate the molarity:

$$\frac{102.9g\ H_2PO_4}{0.642L} \times \frac{mol\ H_3PO_4}{98.0g\ H_3PO_4} = 1.64M\ H_3PO_4$$

Next, calculate the mass-volume percent solution:

$$\frac{102.9g\ H_2PO_4}{642\ mL} \times 100\% = 16.0\%\ mass/volume,\ or\ \left(\frac{16.0g\ H_3PO_4}{100\ mL}\right)$$

Note that the convention in molarity is to divide moles by *liters,* but the convention in mass percent is to divide grams by *milliliters.* If you prefer to think only in terms of liters (not milliliters), then simply consider mass percent as kilograms divided by liters.

---

7. Calculate the molarity of these solutions:

a. 2.0 mol NaCl in 0.872L solution

b. 93g $CuSO_4$ in 390 mL of solution

c. 22g $NaNO_3$ in 777 mL of solution

Solve It

8. How many grams of solute are in each of these solutions?

a. 671 mL of 2.0M NaOH

b. 299 mL of 0.85M HCl

c. 2.74L of 258 mM $Ca(NO_3)_2$

Solve It

9. A 15.0M solution of ammonia, NH_3, has density 0.90g mL^{-1}. What is the mass percent of this solution?

Solve It

10. A chemist dissolves 2.5g of glucose, $C_6H_{12}O_6$, into 375 mL of water. What is the mass percent of this solution? Assuming negligible change in volume upon addition of glucose, what is the molarity of the solution?

Solve It

Changing Concentrations by Making Dilutions

Real-life chemists in real-life labs don't make every solution from scratch. Instead, they make concentrated *stock solutions* and then make *dilutions* of those stocks as necessary for a given experiment.

To make a dilution, you simply add a small quantity of a concentrated stock solution to an amount of pure solvent. The resulting solution contains the amount of solute originally taken from the stock solution, but disperses that solute throughout a greater volume. So, the final concentration is lower; the final solution is less concentrated and more dilute.

 But how do you know how much of the stock solution to use and how much of the pure solvent to use? It depends on the concentration of the stock and on the concentration and volume of the final solution you want. You can answer these kinds of pressing questions by using the dilution equation, which relates concentration (C) and volume (V) between initial and final states:

$$C_1 \times V_1 = C_2 \times V_2$$

 This equation can be used with any units of concentration, provided the same units are used throughout the calculation. Because molarity is such a common way to express concentration, the dilution equation is sometimes expressed in the following way, where M_1 and M_2 refer to the initial and final molarity, respectively:

$$M_1 \times V_1 = M_2 \times V_2$$

EXAMPLE

Q. How would you prepare 500 mL of 200 mM NaOH(aq), given a stock solution of 1.5M NaOH?

A. **Add 67 mL 1.5M NaOH stock solution to 433 mL water.**

Use the dilution equation: $M_1 \times V_1 = M_2 \times V_2$

The initial molarity, M_1, derives from the stock solution, and so is 1.5M. The final molarity is the one you want in your final solution, 200 mM, which is equivalent to 0.200M. The final volume is the one you want for your final solution, 500 mL,

which is equivalent to 0.500L. Using these known values, you can calculate for the initial volume, V_1:

$$V_1 = (0.200M \times 0.500L) / 1.5M = 6.7 \times 10^{-2}L$$

The calculated volume is equivalent to 67 mL. The final volume of the aqueous solution is to be 500 mL. 67 mL of this volume derives from the stock solution. The remainder, 500 mL – 67 mL = 433 mL, derives from pure solvent (water, in this case). So, to prepare the solution, add 67 mL 1.5M stock solution to 433 mL water. Mix and enjoy.

11. What is the final concentration of a solution prepared by diluting 2.50 mL of 3.00M KCl(aq) up to 0.175L final volume?

Solve It

12. A certain mass of ammonium sulfate, $(NH_4)_2SO_4$, is dissolved in water to produce 1.65L of solution. 80.0 mL of this solution is diluted with 120 mL of water to produce 200 mL of 200 mM $(NH_4)_2SO_4$. What mass of ammonium sulfate was originally dissolved?

Solve It

Answers to Questions on Solutions

By this point in the chapter, your brain may feel as if it has itself dissolved. Check your answers, boiling away that confusion to reveal crystalline bits of hard-earned knowledge. In other words, make sure you know what you're doing. Solutions are critically important. Really.

1 **The rank order from most to least soluble is: NaI, NaBr, NaCl, NaF.** As the question indicates, the larger the lattice energy is, the stronger the forces holding together the ions. Dissolving those ions means outcompeting those forces; a solution forms when attractive solute-solvent forces dominate over others (such as solute-solute bonds). So, salts with lower lattice energy are typically more soluble than those with higher lattice energy.

2 **Sugar should be more soluble in ethanol than in octanol.** Like dissolves like. Chemists know from experience that sugar dissolves well in water. Therefore, we expect sugar to dissolve best in solvents that are most similar to water. Because ethanol is more miscible with water than is octanol, we expect that ethanol has solvent properties (especially polarity) more like water than does octanol.

3 The chemist noticed as she swirled the beaker of dissolving cesium sulfate that the beaker was becoming noticeably warmer. This observation momentarily confused her, because it suggested that the dissolution of cesium sulfate released heat, a state of affairs opposite to that usually observed with dissolving salts. Having diagnosed the situation, she cleverly turned it to her advantage. With typical salts, increasing temperature increases solubility in water, so heating a dissolving mixture can promote dissolution. In the case of cesium sulfate, however, the reverse is true: By cooling the dissolving mixture, the chemist promoted solubility of the cesium sulfate.

4 At the high pressures to which deep-sea divers are exposed during their dives, gases become more soluble in the blood and tissue fluids due to Henry's Law (Solubility = Constant × Pressure). So, when the divers do their thing at great depth, high concentrations of these gases dissolve into the blood. If the divers rise to the surface too quickly at the end of a dive, the solubility of these dissolved gases changes too quickly in response to the diminished pressure. This situation can lead to the formation of tiny gas bubbles in the blood and tissues. These bubbles can be deadly.

5 Nothing dramatically fizzy happened when Reefus opened the first soda because that soda was still cold from the refrigerator. As the game progressed, however, the remaining sodas warmed to room temperature as they sat beside Reefus's beanbag chair. Gases (like carbon dioxide) are less soluble in warmer liquids. So, when Reefus opened the warm, fourth-quarter soda, a reservoir of undissolved gas burst forth from the can.

6 4.3×10^{-2} **mol L^{-1}.** To solve this problem, use the two-state form of Henry's Law:

$$S_1 / P_1 = S_2 / P_2$$

The initial solubility and pressure are 0.15 mol L^{-1} and 3.5 atm, respectively. The final pressure is 1.0 atm. Using these known values, solve for the final solubility:

$$S_2 = (0.15 \text{ mol L}^{-1} / 3.5 \text{ atm}) \times 1.0 \text{ atm} = 4.3 \times 10^{-2} \text{ mol L}^{-1}$$

7 Solve these kinds of problems by using the definition of molarity and conversion factors:

a. 2.3M NaCl

$$\frac{2.0 \text{ mol NaCl}}{0.872 \text{L}} = 2.3M \text{ NaCl}$$

b. 1.5_M_ CuSO$_4$

$$\frac{93\text{g CuSO}_4}{390 \text{ mL}} \times \frac{10^3 \text{ mL}}{\text{L}} \times \frac{\text{mol CuSO}_4}{160\text{g CuSO}_4} = 1.5M \text{ CuSO}_4$$

c. 0.33_M_ NaNO$_3$

$$\frac{22\text{g NaNO}_3}{777 \text{ mL}} \times \frac{10^3 \text{ mL}}{\text{L}} \times \frac{\text{mol NaNO}_3}{85.0\text{g NaNO}_3} = 0.33M \text{ NaNO}_3$$

8 Again, conversion factors are the way to approach these kinds of problems. Each problem features a certain volume of solution that contains a certain solute at a certain concentration. Begin each calculation with the given volume. Then convert to moles by multiplying the volume by the concentration. Finally, convert from moles to grams by multiplying by the molar mass of the solute.

a. 54g NaOH

$$\frac{671 \text{ mL}}{1} \times \frac{\text{L}}{10^3 \text{ mL}} \times \frac{2.0 \text{ mol NaOH}}{\text{L}} \times \frac{40.0\text{g NaOH}}{\text{mol NaOH}} = 54\text{g NaOH}$$

b. 9.3g HCl

$$\frac{299 \text{ mL}}{1} \times \frac{\text{L}}{10^3 \text{ mL}} \times \frac{0.85 \text{ mol HCl}}{\text{L}} \times \frac{36.5\text{g HCl}}{\text{mol HCl}} = 9.3\text{g HCl}$$

c. 116g Ca(NO$_3$)$_2$

$$\frac{2.74\text{L}}{1} \times \frac{258 \text{ mmol Ca}\left(\text{NO}_3\right)_2}{\text{L}} \times \frac{\text{mol}}{10^3 \text{ mmol}} \times \frac{164\text{g Ca}\left(\text{NO}_3\right)_2}{\text{mol Ca}\left(\text{NO}_3\right)_2} = 116\text{g Ca}\left(\text{NO}_3\right)_2$$

9 **28%.** To calculate mass percent, you must know the mass of solute and the mass of solution. The molarity of the solution tells you the moles of solute per volume of solution. Starting with this information, you can convert to mass of solute by means of the gram formula mass (see Chapter 7 for details on calculating the gram formula mass):

15.0 mol L^{-1} × 17.0g mol^{-1} = 255g L^{-1}

So, each liter of 15.0_M_ NH$_3$ contains 255g NH$_3$ solute. But how much mass does each liter of solution possess? Calculate the mass of the solution by using the density. Note that the problem lists the density in units of milliliters, so be sure to convert to the proper units:

1.0L solution × (0.90g / 1.0 × 10^{-3}L) = 9.0 × 10^{2}g solution

So, 255g NH$_3$ occur in every 900g of 15.0_M_ NH$_3$. Now you can calculate the mass percent:

Mass percent = 100% × (255g / 900g) = 28%

10 **The mass percent is 0.66%; the molarity is $3.7 \times 10^{-2}M$.** To calculate the mass percent, you must use the estimate that 1.0 mL of water has 1.0g mass. This is a very good approximation at room temperature, and one with which you should be familiar. So, 375 mL water has 375g mass. Adding 2.5g glucose increases that mass to 377.5g for the final solution. Calculate the mass percent as follows:

Mass % = 100% × (2.5g / 377.5g) = 0.66%

To calculate the molarity, you must know the final volume of the solution. Although adding 2.5g to 375 mL water increases the volume from 375 mL, the increase is very small compared to the volume of the water. So, 375 mL is a good approximation of the final volume of the solution. Next, convert from grams of glucose to moles of glucose by means of the gram formula mass:

Moles glucose = 2.5g glucose × (1 mol / 180.2g) = 1.4×10^{-2} mol glucose

Now that you know the moles of glucose and the final volume of solution, calculating molarity is easy:

Molarity = (1.4×10^{-2} mol glucose) / (0.375L solution) = $3.7 \times 10^{-2}M$

11 **$4.29 \times 10^{-2}M$.** Use the dilution equation: $M_1 \times V_1 = M_2 \times V_2$

In this problem, the initial molarity is $3.00M$, the initial volume is 2.50 mL (or 2.50×10^{-3}L), and the final volume is 0.175L. Use these known values to calculate the final molarity, M_2:

$M_2 = (3.00 \text{ mol L}^{-1} \times 2.50 \times 10^{-3}\text{L}) / 0.175\text{L} = 4.29 \times 10^{-2}M$

12 **109g $(NH_4)_2SO_4$.** First, use the dilution equation to find the concentration of the original solution:

$M_1 = (200 \times 10^{-3} \text{ mol L}^{-1} \times 200 \times 10^{-3}\text{L}) / (80.0 \times 10^{-3}\text{L}) = 0.500M$

This calculation means that the original solution contained 0.500 mol $(NH_4)_2SO_4$ per liter of solution. The question indicates that 1.65L of this original solution were prepared, so:

$$\frac{1.65\text{L}}{1} \times \frac{0.500 \text{ mmol } (NH_4)_2 SO_4}{\text{L}} \times \frac{132\text{g}(NH_4)_2 SO_4}{\text{mol }(NH_4)_2 SO_4} = 109\text{g }(NH_4)_2 SO_4$$

Chapter 13

Playing Hot and Cold: Colligative Properties

In This Chapter

▶ Knowing the difference between molarity and molality

▶ Working with boiling point elevations and freezing point depressions

▶ Deducing molecular masses from boiling and freezing point changes

As a recently minted expert in solubility and molarity (see Chapter 12), you may be ready to write off solutions as another chemistry topic mastered, but you, as a chemist worth your salt, must be aware of one final piece to the puzzle. Collectively called the *colligative properties,* these chemically important phenomena arise from the presence of solute particles in a given mass of solvent. The presence of extra particles in a formerly pure solvent has a significant impact on some of that solvent's characteristic properties, such as freezing and boiling points. This chapter walks you through these colligative properties and their consequences and introduces you to a new solution property: molality. No, that's not a typo. Molality is an entirely new property that allows you to solve for the key colligative properties later in this chapter.

Portioning Particles: Molality and Mole Fractions

While Chapter 12 focuses on molarity and its usefulness in expressing the concentration of a solution, this chapter focuses on *molality.* Like the difference in their names, the difference between molarity and molality is subtle. Take a close look at their definitions, expressed next to one another in the following equations:

$$\text{Molarity} = \frac{\text{moles of solute}}{\text{liters of solution}} \quad \text{while} \quad \text{Molality} = \frac{\text{moles of solute}}{\text{kilograms of solvent}}$$

The numerators in molarity and molality calculations are identical, but their denominators differ greatly. Molarity deals with liters of solution, while molality deals with kilograms of solvent. In Chapter 12, you find out that a solution is a mixture of solvent and solute; a solvent is the medium into which the solute is mixed.

A further subtlety to the molarity/molality confusion is how to distinguish between their variables. The letter "m" turns out to be an overused variable in chemistry. By the time chemists got to molality, they had already defined mass, molarity, and the mole and were fresh out of "m's" to use. Instead of picking another variable (or perhaps a less confusing name that started with a nice uncommon letter like z), chemists decided to give molality the "script" m as its variable (m). To help you avoid uttering any four-letter words when confronted with this plethora of m-words and their variables, we have provided you with Table 13-1 containing all of them.

Table 13-1	The Chemical M Words
Name	*Abbreviation*
Mass	m
Molarity	*M*
Molality	*m*
Moles	mol

Be very careful of the wording of a molality problem when the solute is an ionic compound. As we explain in Chapter 12, ionic compounds dissociate in aqueous solutions, splitting up into their constituent particles. Because colligative properties depend on the total number of particles *in solution* and not the total number of particles of solute *before* it's added to the solution, you need to take this property into account. For example, one mole of iron (III) chloride ($FeCl_3$) dissociates into four moles of particles in water, because water splits each molecule of iron (III) chloride into three chlorine ions and one iron ion.

You may also occasionally be asked to calculate something called the *mole fraction* of a solution, which is the ratio of the number of moles of *either* solute or solvent in a solution to the total number of moles of solute *and* solvent in the solution. By the time chemists defined this quantity, however, they had finally acknowledged that they had too many "m" variables, and they gave it the variable X. Of course, chemists still need to distinguish between the mole fraction of the solute and the solvent, which unfortunately both start with the letter "s." To avoid further confusion, they decided to abbreviate solute and solvent as A and B respectively in the general formula, although in practice, the chemical formulas of the solute and solvent are usually written as subscripts in place of A and B. For example, the mole fraction of sodium chloride in a compound would be written as X_{NaCl}.

In general, the mole ratio of the solute in a solution is expressed as

$$X_A = \frac{n_A}{n_A + n_B}$$

Where n_A is the number of moles of solute and n_B is the number of moles of solvent. The mole ratio of the solvent is then

$$X_B = \frac{n_B}{n_A + n_B}$$

These mole fractions are useful because they represent the ratio of solute to solution and solvent to solution very well and give you a general understanding of how much of your solution is solute and how much is solvent.

EXAMPLE

Q. How many grams of sodium chloride must you add to 750g of water to make a 0.35 molal solution?

A. **7.59g NaCl.** This problem gives you molality and the mass of a solvent and asks you to solve for the mass of solute. Because molality involves moles and not grams of solute, you need to first solve for moles of solvent, and then use the gram formula mass of sodium chloride to solve for the number of grams of solute. Before plugging the numbers into the molality equation, you must also note that the problem has given you the mass of the solvent in grams, while the formula calls for it to be in kilograms. If you have been moving through this book sequentially, you should be familiar enough with unit conversions to know that moving from grams to kilograms is equivalent to moving the decimal point to the left three places, but if you need a refresher on the ins and outs of unit conversions, please refer to Chapter 2. Plugging everything you know into the equation for molality gives you

$$0.35m = \frac{X \text{ mol NaCl}}{0.750 \text{ kg H}_2\text{O}}$$

Solving for the unknown gives you 0.26 mol NaCl in solution. Recall, however, that NaCl is an ionic compound that dissociates into two moles of solute for every mole of dry NaCl added to the solution, so this number is twice as much as the number of moles of NaCl added originally (0.13 mol). To finish up the problem, you must convert this measurement to grams using the gram formula mass.

$$\frac{0.13 \text{ mol NaCl}}{1} \times \frac{58.4\text{g}}{1 \text{ mol NaCl}} = 7.59\text{g NaCl}$$

---

1. 50g of potassium iodide is added to 1,500g of water. What is the molality of the resulting solution?

Solve It

2. How many grams of magnesium fluoride must you add to 3,200g of water to make a 0.42 molal solution?

Solve It

3. Calculate the mole fraction of each component in a solution containing 2.75 mol ethanol and 6.25 mol water.

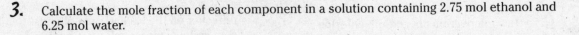

Solve It

Too Hot to Handle: Elevating and Calculating Boiling Points

Calculating molality is no more or less difficult than calculating molarity, so you may be asking yourself "why all the fuss?" Is it even worth adding another quantity and another variable to memorize? Yes! While molarity is exceptionally convenient for calculating concentrations and working out how to make dilutions in the most efficient way, molality is reserved for the calculation of several important colligative properties, including *boiling point elevation*. Boiling point elevation refers to the tendency of a solvent's boiling point to increase when an impurity (a solute) is added to it. In fact, the more solute that is added, the greater the change in the boiling point.

Boiling point elevations are directly proportional to the molality of a solution, but chemists have found that some solvents are more susceptible to this change than others. The formula for the change in the boiling point of a solution, therefore, contains a proportionality constant, abbreviated K_b, which is a property determined experimentally and must be read from a table such as Table 13-2. The formula for the boiling point elevation is

$$\Delta T_b = K_b \times m$$

Note the use of the Greek letter delta (Δ) in the formula to indicate that you're calculating a *change in* boiling point, not the boiling point itself. You'll need to add this number to the boiling point of the pure solvent to get the boiling point of the solution. The units of K_b are typically given in degrees Celsius per molality.

Table 13-2	Common K_b Values	
Solvent	*K_b in °C/m*	*Boiling Point in °C*
Acetic acid	3.07	118.1
Benzene	2.53	80.1
Camphor	5.95	204.0
Carbon tetrachloride	4.95	76.7
Cyclohexane	2.79	80.7
Ethanol	1.19	78.4
Phenol	3.56	181.7
Water	0.512	100.0

Boiling point elevations are a result of the attraction between solvent and solute particles in a solution. Adding solute particles increases these intermolecular attractions because more particles are around to attract one another. Solvent particles must therefore achieve a greater kinetic energy to overcome this extra attractive force and boil, which translates into a higher boiling point. (See Chapter 10 for full information on kinetic energy.)

Q. What is the boiling point of a solution containing 45.2g of menthol ($C_{10}H_{20}O$) dissolved in 350g of acetic acid?

A. **120.7°C.** The problem asks for the boiling point of the solution, so you know that first you have to calculate the boiling point elevation. This means you're required to know the molality of the solution and the K_b value of the solvent (acetic acid). Table 13-2 tells you that the K_b of acetic acid is 3.07. To calculate the molality, you must convert 45.2g of menthol to moles.

$$\frac{45.2\text{g menthol}}{1} \times \frac{1 \text{ mol menthol}}{156\text{g menthol}} = 0.29 \text{ mol}$$

You can now calculate the molality of the solution, taking care to convert grams of acetic acid to kilograms.

$$m = \frac{0.29 \text{ mol menthol}}{0.350 \text{ kg acetic acid}} = 0.83$$

Now that you have molality, you can plug it and your K_b into the formula to find the change in boiling point.

$$\Delta T_b = 3.07°C/m \times 0.83m = 2.5°C$$

Remember, however, that you're not quite done because the problem asks for the boiling point of the solution, not the change in it. Luckily, the last step is just simple arithmetic. You must add your ΔT_b to the boiling point of pure acetic acid which, according to Table 13-2, is 118.1°C. This gives you a final boiling point for the solution of 118.1°C + 2.6°C = 120.7°C.

4. What is the boiling point of a solution containing 158g sodium chloride (NaCl) and 1.2 kg of water? What if the same number of moles of calcium chloride ($CaCl_2$) is added to the solvent? Explain why there is such a great difference in the boiling point elevation?

Solve It

5. A clumsy chemist topples a bottle of indigo dye ($C_{16}H_{10}N_2O_2$) into a beaker containing 450g of ethanol. If the boiling point of the resulting solution is 79.2°C, how many grams of dye were in the bottle?

Solve It

How Low Can You Go? Depressing and Calculating Freezing Points

The second of the important colligative properties that can be calculated from molality is *freezing point depression*. Adding solute to a solvent not only raises its boiling point; it also lowers its freezing point. This is the reason, for example, that you sprinkle salt on icy sidewalks. The salt mixes with the ice and lowers its freezing point. If this new freezing point is lower than the outside temperature, the ice melts, eliminating the spectacular wipeouts so common on salt-free sidewalks. The colder it is outside, the more salt is needed to melt the ice and lower the freezing point to below the ambient temperature.

Freezing point depressions, like boiling point elevations, are calculated using a constant of proportionality, this time abbreviated K_f. The formula therefore becomes $\Delta T_f = K_f \times m$. To calculate the new freezing point of a compound, you must *subtract* the change in freezing point from the freezing point of the pure solvent. Table 13-3 lists several common K_f values.

Table 13-3	Common K_f Values	
Solvent	*K_f in m/°C*	*Freezing Point in °C*
Acetic acid	3.90	16.6
Benzene	5.12	5.5
Camphor	37.7	179
Carbon tetrachloride	30.0	−23
Cyclohexane	20.2	6.4
Ethanol	1.99	−114.6
Phenol	7.40	41
Water	1.86	0.0

Adding an impurity to a solvent makes its liquid phase more stable through the combined effects of boiling point elevation and freezing point depression. This is the reason why you rarely see bodies of frozen salt water. The salt in the oceans lowers the freezing point of the water, making the liquid phase more stable and able to sustain temperatures slightly below 0°C.

Q. Each kilogram of seawater contains roughly 35g of dissolved salts. Assuming that all of these salts are sodium chloride, what is the freezing point of seawater?

A. **−2.23°C.** Here again, you need to begin by converting grams of salt to moles to figure out the molality. One mole of NaCl is equivalent to 58.4g, so 35g is equivalent to 0.60 mol of NaCl. This number must be multiplied by 2 to compensate for the fact that sodium chloride dissociates into twice as many particles in water, so this solution contains 1.20 mol. Next, you must find the molality of the solution by dividing this number of moles by the mass of the solvent (1 kg),

giving a 1.20 molal solution. Lastly, you must look up the K_f of water in Table 13-3 and plug all of these values into the equation for freezing point depression, giving you

$$\Delta T_f = 1.86°C/m \times 1.20m = 2.23°C$$

Because this is merely the freezing point depression, you must subtract it from the freezing point of the pure solute to get $0 - 2.23°C = -2.23°C$, the freezing point of seawater.

6. Antifreeze takes advantage of freezing point depression to lower the freezing point of the water in your car's engine and keep it from freezing on blistery winter drives. If antifreeze is made primarily of ethylene glycol ($C_2H_6O_2$), how much of it needs to be added to lower the freezing point of 10.0 kg of water by 15.0°C?

Solve It

7. If 15g of silver is dissolved into 1,500g of ethanol, what is the freezing point of the mixture?

Solve It

Determining Molecular Masses with Boiling and Freezing Points

Just as a solid understanding of molality helps you to calculate changes in boiling and freezing points, a solid understanding of ΔT_b and ΔT_f can help you determine the molecular mass of a mystery compound being added to a known quantity of solvent. When you're asked to solve problems of this type, you'll always be given the mass of the mystery solute, the mass of solvent, and either the change in the freezing or boiling point or the new freezing or boiling point itself. From this information you then follow a set of simple steps to determine the molecular mass.

1. If you've been given the boiling point, calculate the ΔT_b by subtracting the boiling point of the pure solvent from the number you were given. If you know the freezing point, add the freezing point of the pure solvent to it to get the ΔT_f.

2. Look up the K_b or K_f of the solvent (refer back to Tables 13-2 and 13-3).

3. Solve for the molality of the solution using the equation for ΔT_b or ΔT_f.

4. Calculate the number of moles of solute in the solution by multiplying the molality calculated in Step 3 by the number of kg of solvent you were given at the start of the problem.

5. Divide the mass of solute you were given originally by the number of moles calculated in Step 4. This is your molecular mass, or number of grams per mole, from which you can often guess the identity of the mystery compound.

EXAMPLE

Q. 97.3g of a mystery compound is added to 500g of water, raising its boiling point to 100.78°C. What is the molecular mass of the mystery compound?

A. **128.0 g/mol.** According to the preceding steps, you must first subtract the boiling point of water from this new boiling point, giving you

$$\Delta T_b = 100.78°C - 100.00°C = 0.78°C$$

You then plug this value and a K_b of 0.512 into the equation for boiling point elevation and solve for molality, giving you

$$m = \frac{\Delta T_b}{K_b} = \frac{0.78°C}{0.512°C/m} = 1.52$$

Next, take this molality value and multiply it by the provided value for mass of solvent, giving you

$$\frac{1.52 \text{ mol solute}}{1 \text{ kg H}_2\text{O}} \times 0.5 \text{ kg H}_2\text{O} = 0.76 \text{ mol solute}$$

Lastly, you must divide the number of grams of the mystery solute by the number of moles, giving the molecular mass of the compound

$$\frac{97.3\text{g}}{0.76 \text{ mol}} = 128.0 \text{ g/mol}$$

---

8. The freezing point of 83.2g of carbon tetrachloride is lowered by 11.52°C when 15.0g of a mystery compound is added to it. What is the molecular mass of this mystery compound?

Solve It

9. Benzene's boiling point of 42.1g is raised to 81.9°C when 8.8g of a mystery compound is added to it. What is the molecular mass of this mystery compound?

Solve It

Answers to Questions on Colligative Properties

You've tried your hand at problems on molality, boiling points, and freezing points. Are you feeling a little hot under the collar or as cool as a cucumber? Here, we present the answers to the practice problems for this chapter.

1 **0.40m.** To calculate the molality of this potassium iodide solution, you must first find out how many moles of potassium iodide are in the solution. This requires knowing its chemical formula, its gram formula mass, and how many constituent particles it will dissociate into in solution. Potassium is an alkali metal with a typical ionic charge of +1, while iodine is a halogen with a typical ionic charge of –1. These two elements therefore combine in a one-to-one ratio, giving the chemical formula KI (see Chapter 6 for more about naming compounds). Adding their mass numbers gives a gram formula mass of 166.0g per mole. Dividing the given mass of potassium iodide (50g) by this number tells you that 0.30 mol potassium iodide was added to the solution. Because potassium iodide dissociates into two particles per mole of dry solute, multiply this number by 2 to give 0.60 mol of solute in solution. Divide this number by the mass of water in kilograms to give

$$m = \frac{0.60 \text{ mol KI}}{1.5 \text{ kg H}_2\text{O}} = 0.40m$$

2 **1.08 kg MgI$_2$.** This problem gives you the molality and asks you to solve for the solute amount. Plugging all the known variables into the molality equation gives

$$0.42m = \frac{X \text{ mol MgI}_2}{3.2 \text{ kg H}_2\text{O}}$$

Solving this equation for the unknown number of moles of solute gives you 1.3 mol MgI$_2$. Because magnesium iodide dissociates into three particles per mole of dry solute in solution, multiply this number by 3 to give 3.9 mol MgI$_2$ in solution. The final step is to convert this to a mass in grams by using the gram formula mass of magnesium iodide (278g per mole), giving you 1,084g MgI$_2$, or 1.08 kg.

3 **The solution is 31% ethanol and 69% water.** This problem tells you that n_{EtOH} is 2.75 mol and n_{H2O} is 6.25 mol. Plugging these into the equations for the mole fraction of solute and solvent yields

$$X_{EtOH} = \frac{2.75 \text{ mol}}{2.75 \text{ mol} + 6.25 \text{ mol}} = 0.31 \text{ and } X_{H_2O} = \frac{6.25 \text{ mol}}{2.75 \text{ mol} + 6.25 \text{ mol}} = 0.69$$

4 The solution containing sodium chloride has a boiling point of **102.3°C;** the solution containing calcium chloride has a boiling point of **103.5°C.** To solve for the boiling point, you must first solve for the molality. Start by dividing 158g NaCl by its gram formula mass (58.4g per mole), which tells you that there are 2.71 mol of solute before it's added to the solution. Multiply this by 2 because each molecule of NaCl splits into two particles in solution, for a total of 5.42 mol. Divide this by the mass of solvent (1.2 kg) to give you a molality of 4.5m. Finally, multiply this molality by the K_b of water, 0.512°C/m, to give you a ΔT_b of 2.3°C. Add it to the boiling point of pure water (100°C) to give you a new boiling point of 102.3°C.

If the same number of moles of calcium chloride ($CaCl_2$) were added, on the other hand, it would dissociate into three particles per mole in solution, giving 2.71 mol × 3 = 8.13 mol of solute in solution. As with the sodium chloride solution, divide this by the mass of solvent (1.2 kg) to get 6.8m, and multiply by the K_b of water (0.512°C/m) to get a ΔT_b of 3.5°C. This is a difference of more than 1 degree! The difference arises because colligative properties such as boiling point elevation depend only on the number of particles *in solution*.

5   **79.3g $C_{16}H_{10}N_2O_2$.** This problem requires a little bit more thought than the others you've seen before because it requires you to solve a boiling point elevation problem backwards. You're given the boiling point of the solution, so the first thing to do is to solve for the ΔT_b of the solution. You do this by subtracting the boiling point of pure ethanol from the given boiling point of the impure solution. This gives you

ΔT_b= 79.2°C – 78.4°C = 0.8°C

After looking up the K_b of ethanol (1.19°C/m), the only unknown in your ΔT_b equation is the molality. Solving for this gives you

$$m = \frac{\Delta T_b}{K_b} = \frac{0.8°C}{1.19°C/m} = 0.67m$$

Now that you know the molality, you can extract the number of moles of solute.

$$0.67m = \frac{X \text{ mol } C_{16}H_{10}N_2O_2}{0.450 \text{ kg EtOH}} \Rightarrow 0.30 \text{ mol } C_{16}H_{10}N_2O_2$$

Lastly, translate this mole count into a mass by multiplying by the gram molecular mass of $C_{16}H_{10}N_2O_2$ (262g per mole), giving you your final answer of 79.3g $C_{16}H_{10}N_2O_2$.

6   **5.00 kg $C_2H_6O_2$.** Here you've been given a freezing point depression of 15.0°C and have been asked to backsolve for the number of grams of antifreeze required to make it happen. Begin by solving for the molality of the solution by plugging all the known quantities into your freezing point depression equation and solving for molality.

$$m = \frac{\Delta T_f}{K_f} = \frac{15.0°C}{1.86°C/m} = 8.06m$$

Now solve for the number of moles of the solvent ethylene glycol (antifreeze).

$$8.06m = \frac{X \text{ mol } C_2H_6O_2}{10.0 \text{ kg } H_2O} \Rightarrow 80.6 \text{ mol } C_2H_6O_2$$

Finally, translate this mole count into a mass by multiplying by the gram molecular mass of $C_2H_6O_2$ (62.0g per mole), giving you your final answer of 5.00 kg $C_2H_6O_2$. Lowering the freezing point of water by such a significant amount requires a solution that is one-third antifreeze by mass!

7   **–114.4°C.** First calculate the molality of the solution by converting the mass of silver into a mole count (15g × (1 mol/108g) = 0.14 mol) and dividing it by the mass of ethanol in kilograms, giving you 0.093m. Then multiply this molality by the K_f of ethanol (1.99°C/m), giving you a ΔT_f

of 0.19°C. Lastly, subtract this ΔT_f value from the freezing point of pure ethanol, giving you

$-114.6°C - 0.19°C = -114.4°C$

8 **470 g/mol.** Follow the steps outlined in the "Determining Molecular Masses with Boiling and Freezing Points" section carefully to arrive at the correct answer. Step 1 is unnecessary in this case because you've already been given the ΔT_f and don't need to solve for it, so begin by looking up the K_f of carbon tetrachloride in Table 13-3 ($30.0°C/m$). Plug both of these values into the equation for boiling point elevation and solve for molality, giving you

$$m = \frac{\Delta T_f}{K_f} = \frac{11.52°C}{30.0°C/m} = 0.384m$$

Next, take this molality value and multiply it by the provided value for mass of solvent, giving

$$\frac{0.384 \text{ mol solute}}{1 \text{ kg H}_2\text{O}} \times 0.083 \text{ kg} = 0.0319 \text{ mol solute}$$

Lastly, you must divide the number of grams of the mystery solute by the number of moles, giving the molecular mass of the compound

$$\frac{15g}{0.0319 \text{ mol}} = 470 \text{ g/mol}$$

9 **293 g/mol.** Unlike in Question 8, here you can't ignore the first step in the procedure because you aren't given the ΔT_b directly. Solve for it by subtracting the boiling point of pure benzene (refer to Table 13-2) from the given boiling point of the solution, giving you

$81.9° - 80.1°C = 1.80°C$

Plug this and the K_b for benzene ($2.53°C/m$) into your ΔT_b and solve for the molality of the solution.

$$m = \frac{\Delta T_b}{K_b} = \frac{1.80°C}{2.53°C/m} = 0.71m$$

Next, take this molality value and multiply it by the provided value for mass of solvent, giving you

$$\frac{0.71 \text{ mol solute}}{1 \text{ kg H}_2\text{O}} \times 0.0421 \text{ kg} = 0.030 \text{ mol solute}$$

Lastly, you must divide the number of grams of the mystery solute by the number of moles, giving you the molecular mass of the compound.

$$\frac{8.8g}{0.030 \text{ mol}} = 293 \text{ g/mol}$$

Chapter 14

Exploring Rate and Equilibrium

· ·

In This Chapter

▶ Measuring rates and understanding the factors that affect them

▶ Measuring equilibrium and understanding how it responds to disruption

· ·

Most people don't like waiting. And nobody likes waiting for nothing. Research has tentatively concluded that chemists are people, too. It follows that chemists don't like to wait, and if they must wait, they'd prefer to get something for their trouble.

To address these concerns, chemists study things like *rate* and *equilibrium*.

✔ **Rates** tell chemists how long they'll have to wait for a reaction to occur.

✔ **Equilibrium** tells chemists how much product they'll get if they wait long enough.

These two concepts are completely separate. There's no connection between how productive a reaction can be and how long it takes for that reaction to proceed. In other words, chemists have good days and bad days, like everyone else. At least they have a little bit of theory to help them make sense of these things. In this chapter, you get an overview of this theory. Don't wait. Read on.

Measuring Rates

So, you've got this beaker, and a reaction is going on inside of it. Is the reaction a fast one or a slow one? How fast or how slow? How can you tell? These are questions about rates. You can measure a reaction rate by measuring how fast a reactant disappears or by measuring how fast a product appears. If the reaction occurs in solution, the molar concentration of reactant or product changes over time, so rates are often expressed in units of molarity per second ($M\,s^{-1}$).

For the following reaction,

$$A + B \rightarrow C$$

you can measure the reaction rate by measuring the decrease in the concentration of either reactant A (or B) or the increase in the concentration of product C over time:

$$\text{Rate} = -d[A] / dt = d[C] / dt$$

In these kinds of equations, d is math-speak for a change in the amount of something at any given moment. If you plot the concentration of product against the reaction time, for example, you might get a curve like the one shown in Figure 14-1. Reactions usually occur most quickly at the beginning of a reaction, when the concentration of products is the lowest and the concentration of reactants is highest. The precise rate at any given moment of the reaction is called the *instantaneous rate* and is equal to the slope of a line drawn tangent to the curve.

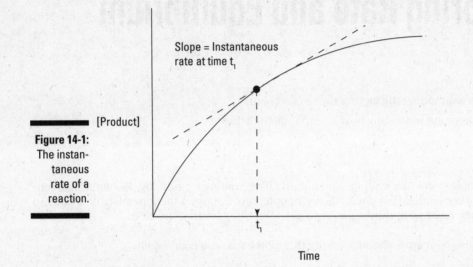

Figure 14-1:
The instantaneous rate of a reaction.

Slope = Instantaneous rate at time t_1

[Product]

t_1

Time

Clearly, the rate can change as the concentration of reactants and/or products changes. So, you'd expect that any equation for describing the rate of a reaction like the one shown in Figure 14-1 would include some variable for the concentration of reactant. You'd be right.

Equations that relate the rate of a reaction to the concentration of some *species* (which can mean either reactant or product) in solution are called *rate laws*. The exact form a rate law assumes depends on the reaction involved. Countless research studies have described the intricacies of rate in chemical reactions. Here, we focus on rate laws for simpler reactions. In general, rate laws take the form:

Rate = k [reactant A]^m[reactant B]^n

The rate law shown describes a reaction whose rate depends on the concentration of two reactants, A and B. Other rate laws for other reactions may include factors for greater or fewer reactants. In this equation, k is the *rate constant,* a number that must be experimentally measured for different reactions. The exponents m and n are called *reaction orders,* and must also be measured for different reactions. A reaction order reflects the impact of a change in concentration in overall rate. If $m > n$, then a change in the concentration of A affects the rate more than does changing the concentration of B. The sum, $m + n$, is the overall reaction order.

Some simple kinds of reaction rate laws crop up frequently, so they're worth your notice:

✔ **Zero-order reactions:** Rates for these reactions don't depend on the concentration of any species, but simply proceed at a characteristic rate:

Rate = k

✔ **First-order reactions:** Rates for these reactions typically depend on the concentration of a single species:

> Rate = $k[A]$

✔ **Second-order reactions:** Rates for these reactions may depend on the concentration of two species, or may have second-order dependence on the concentration of a single species (or some intermediate combination):

> Rate = $k[A][B]$ or
>
> Rate = $k[A]^2$

Measuring rates can help you figure out *mechanism,* the molecular details by which a reaction takes place. The slowest step is typically the one that contributes most to the observed rate — it's the *rate determining step.* By measuring reaction rates under varying conditions, you can discover a lot about the chemical nature of the rate determining step. Be forewarned, however: You can never use rates to prove a reaction mechanism; you can only use rates to disprove incorrect mechanisms.

EXAMPLE

Q. Consider the following two reactions and their associated rate laws:

Reaction 1: $A + B \rightarrow C$
Rate = $k[A]^2$

Reaction 2: $D + E \rightarrow F + 2G$
Rate = $k[D][E]$

a. What is the overall reaction order for Reaction 1 and for Reaction 2?

b. For Reaction 1, how will the rate change if the concentration of A is doubled? How will the rate change if the concentration of B is doubled?

c. For Reaction 2, how will the rate change if the concentration of D is doubled? How will the rate change if the concentration of E is doubled?

d. What is the relationship between the rates of change in [A], [B], and [C]? What is the relationship between the rates of change in [D], [E], [F], and [G]?

A. Based on the given reactions and their associated rate laws, here are answers to the questions about reaction order and rate:

a. Both reactions are second order, because adding the individual reaction orders yields "2" in each case.

b. If [A] is doubled, the rate law predicts that the rate will quadruple, because $2^2 = 4$. If [B] is doubled, the rate won't change because [B] doesn't appear in the rate law.

c. If [D] is doubled or if [E] is doubled, the rate law predicts that the rate will also double because $2^1 = 2$.

d. Rate = $d[A] / dt = d[B] / dt = -d[C] / dt$
and

Rate = $d[D] / dt = d[E] / dt = -d[F] / dt = -0.5d[G] / dt$

In Reaction 1, reactants A and B are consumed to make product C, so the rates of change in the concentration of A and B will have the opposite sign as the rate of change in C (you could swap positive and negative values throughout the rate equation, and the answer would still be correct). In Reaction 2, the same guidelines apply for positive/negative signs, but there's an added wrinkle: 2 moles of product G are made for every 1 mole made of product F, and for every 1 mole consumed of reactants D and E. So, the rate of change in D, E, and F is one half the rate of change in G, as indicated by the coefficient of 0.5.

1. Methane combusts with oxygen to yield carbon dioxide and water vapor:

$$CH_4 + 2O_2 \rightarrow CO_2 + 2H_2O$$

If methane is consumed at 2.79 mol s^{-1}, what is the rate of production of carbon dioxide and water?

2. You study the following reaction: $A + B \rightarrow C$

You observe that tripling the concentration of A increases the rate by a factor of 9. You also observe that doubling the concentration of B doubles the rate. Write the rate law for this reaction. What is the overall reaction order?

3. You study the following reaction: $D + E \rightarrow F + 2G$

You vary the concentration of reactants D and E, and observe the resulting rates:

	[D], M	[E], M	Rate, M s^{-1}
Trial 1:	2.7×10^{-2}	2.7×10^{-2}	4.8×10^{6}
Trial 2:	2.7×10^{-2}	5.4×10^{-2}	9.6×10^{6}
Trial 3:	5.4×10^{-2}	2.7×10^{-2}	9.6×10^{6}

Write the rate law for this reaction and calculate the rate constant, k. At what rate will the reaction occur in the presence of $1.3 \times 10^{-2} M$ reactant D and $0.92 \times 10^{-2} M$ reactant E?

Solve It

Focusing on Factors that Affect Rates

Despite the impression given by their choice in clothing, chemists are finicky, tinkering types. They usually want to change reaction rates to suit their own needs. What can affect rates, and why? Temperature, concentration, and catalysts influence rate as follows:

- **Reaction rates tend to increase with temperature.** This trend results from the fact that reactants must collide with one another to have the chance to react. If reactants collide with the right orientation and with enough energy, the reaction can occur. So, the greater the number of collisions, and the greater the energy of those collisions, the more actual reacting takes place. An increase in temperature corresponds to an increase in the average kinetic energy of the particles in a reacting mixture — the particles move faster, colliding more frequently and with greater energy. (See Chapter 10 for more about kinetic energy.)

- **Increasing concentration tends to increase reaction rate.** The reason for this trend also has to do with collisions. Higher concentrations mean the reactant particles are closer to one another, so they undergo more collisions and have a greater chance of reacting. Increasing the concentration of reactants may mean dissolving more of those reactants in solution. Some reactants aren't completely dissolved, but come in larger, undissolved particles. In these cases, smaller particles lead to faster reactions. Smaller particles expose more surface area, making a greater portion of the particle available for reaction.

- **Catalysts increase reaction rates.** Catalysts don't themselves become chemically changed, and they don't alter the amount of product a reaction can eventually produce (the *yield*). Catalysts can operate in many different ways, but all of those ways have to do with decreasing *activation energy,* the energetic hill reactants must climb to reach a *transition state,* the highest-energy state along a reaction pathway. Lower activation energies mean faster reactions. Figure 14-2 shows a *reaction progress diagram,* the energetic pathway reactants must traverse to become products. The figure also shows how a catalyst affects reactions, by lowering the activation energy without altering the energies of the reactants or products.

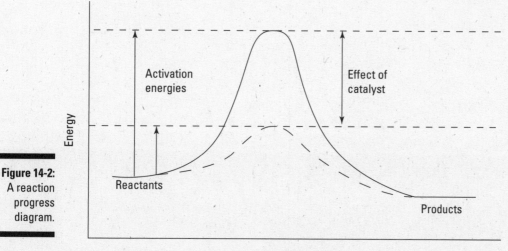

Figure 14-2:
A reaction progress diagram.

Q. Consider the following reaction:
$H_2(g) + Cl_2(g) \rightarrow 2HCl(g)$

If 1 mol H_2 reacts with 1 mol Cl_2 to form 2 mol HCl, does the reaction occur more rapidly in a 5L or a 10L vessel? Does it occur more rapidly at 273K or 293K? Why?

A. The reaction occurs more rapidly in the 5L vessel because the concentration of reactant molecules is higher when they occupy the smaller volume. At higher concentration, more collisions occur between reactant molecules. At higher temperatures, particles move with greater energy, which also produces more collisions, and collisions of greater force. So, the reaction occurs more rapidly at 293K.

---

4. Here is a simplified reaction equation for the combustion of gunpowder:

$$10KNO_3 + 3S + 8C \rightarrow 2K_2CO_3 + 3K_2SO_4 + 6CO_2 + 5N_2$$

In the 1300s, powdersmiths began to process raw gunpowder by using a method called *corning*. Corning involves adding liquid to gunpowder to make a paste, pressing the paste into solid cakes, and forcing the cakes through a sieve to produce grains of defined size. These grains are both larger and more consistently sized than the original grains of powder. Explain the advantages of corned gunpowder from the perspective of chemical kinetics.

Solve It

5. Consider the following reaction: $A \rightarrow B$

To progress from reactant to product, reactant A must pass through a high-energy transition state, A*. Imagine that reaction conditions are changed in such a way that the energy of A is increased, the energy of B is decreased, and the energy of A* remains unchanged. Does the change in conditions result in a faster reaction? Why or why not?

Solve It

Measuring Equilibrium

Figure 14-3 shows a reaction progress diagram like the one shown in Figure 14-2, but highlights the difference in energy between reactants and products. This difference is completely independent of activation energy, which we discuss in the previous section. Although activation energy controls the rate of a reaction, the difference in energy between reactants and products determines the extent of a reaction — how much reactant will have converted to product when the reaction is complete.

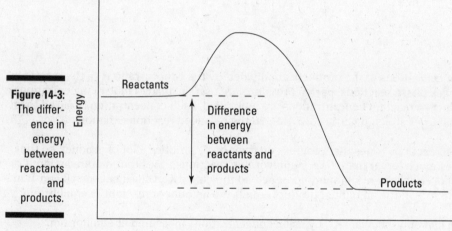

Figure 14-3:
The difference in energy between reactants and products.

Reaction progress

A reaction that has produced as much product as it ever will is said to be at *equilibrium*. Equilibrium doesn't mean that no more chemistry is occurring; rather, equilibrium means that the rate at which reactant converts to product equals the rate at which product converts back into reactant. So, the overall concentrations of reactants and products no longer change.

Two key ideas emerge:

✔ First, the equilibrium position of a reaction (the extent to which the reaction proceeds) can be characterized by the concentrations of reactants and products, because these concentrations no longer change.

✔ Second, the equilibrium position of a reaction is intimately connected to the difference in energy between reactants and products.

These two ideas are expressed by two important equations, which we cover in the upcoming sections.

The equilibrium constant

First, a parameter called K_{eq}, the *equilibrium constant*, describes the equilibrium position of a reaction in terms of the concentration of reactants and products. For the following reaction,

$$aA + bB \leftrightarrow cC + dD$$

The equilibrium constant is calculated as follows:

$$K_{eq} = \frac{[C]^c [D]^d}{[A]^a [B]^b}$$

In general, concentrations of the products are divided by the concentrations of the reactants. In the case of gas phase reactions, partial pressures are used instead of molar concentrations. Multiple product or reactant concentrations are multiplied. Each concentration is raised to an exponent equal to its stoichiometric coefficient in the balanced reaction equation.

Very favorable reactions (ones that progress on their own) produce a lot of product, so they have K_{eq} values much larger than 1. Very unfavorable reactions (ones that require an input of energy) convert very little reactant into product, so they have K_{eq} values between 0 and 1. In a reaction with $K_{eq} = 1$, the amount of product equals the amount of reactant at equilibrium.

Note that you can only calculate K_{eq} by using concentrations measured at equilibrium. Concentrations measured before a reaction reaches equilibrium are used to calculate a *reaction quotient, Q:*

$$Q = \frac{[C]^c [D]^d}{[A]^a [B]^b}$$

If $Q < K_{eq}$, the reaction will progress "to the right," making more product. If $Q > K_{eq}$, the reaction will shift "to the left," converting product into reactant. If $Q = K_{eq}$, the reaction is at equilibrium.

Free energy

The second key equation having to do with equilibrium relates the K_{eq} to the difference in energy between reactants and products. The particular form of energy important in this relationship is *free energy, G*. The difference in free energy between product and reactant states is $\Delta G = G_{product} - G_{reactant}$. The relationship between free energy and equilibrium is

$$\Delta G = -RT \ln K_{eq}$$

or

$$K_{eq} = e^{-\Delta G / RT}$$

In these equations, R is the gas constant (0.08206L atm K^{-1} mol^{-1} or 8.314L kPa K^{-1} mol^{-1}; see Chapter 11 for more information), T is temperature, and ln refers to the natural logarithm, log base e. The equation is typically true for reactions that occur with no change in temperature or pressure.

Favorable reactions possess negative values for ΔG and unfavorable reactions possess positive values for ΔG. Energy must be added to drive an unfavorable reaction forward. If the ΔG for a set of reaction conditions is 0, the reaction is at equilibrium.

The free energy change for a reaction, ΔG, arises from the interplay of two other parameters, the *enthalpy* change, ΔH, and the *entropy* change, ΔS. A detailed discussion of enthalpy and entropy is well beyond the scope of this book. As a *rough* approximation, you can think of enthalpy as the energy associated with order in the system, and entropy as the energy associated with disorder in the system. Negative changes in enthalpy are favorable, as are positive changes in entropy. Overall, then:

$\Delta G = \Delta H - T\Delta S$ (where T is temperature)

EXAMPLE

Q. Consider the following reaction:
A + 2B $\leftrightarrow$ 2C

a. Write the expression that relates the equilibrium constant for this reaction to the concentrations of the reactants and product.

b. If the $K_{eq} = 1.37 \times 10^3$, what is the free energy change for the reaction at 298K?

c. At equilibrium, are reactants or product favored?

(Note: $R = 8.314$L kPa K^{-1} mol^{-1} = 8.314J K^{-1} mol^{-1})

A. Here are answers for the questions about the reaction: A + 2B $\leftrightarrow$ 2C

a. $K_{eq} = [C]^2 / ([A][B]^2)$

b. $\Delta G = -RT \ln K_{eq} =$
$-(8.314$J K^{-1} mol$^{-1})(298$K$) \ln 1.37 \times 10^3 =$
-1.79×10^4J mol^{-1} =
-17.9 kJ mol^{-1}

c. $K_{eq} > 1$ and $\Delta G < 0$, so product is favored.

6. Consider the following reaction:
$N_2(g) + O_2(g) \leftrightarrow 2NO(g)$

At equilibrium, you measure the following partial pressures: $P(N_2) = 4.76 \times 10^{-2}$ atm, $P(O_2) = 9.82 \times 10^{-3}$ atm, and $P(NO) = 2.63 \times 10^{-7}$ atm.

a. What is the K_{eq} for this reaction?

b. If you measured $P(O_2) = 3.74 \times 10^{-2}$ atm, and other partial pressures remained unchanged, what is the reaction quotient for the reaction? In which direction would the reaction proceed?

Solve It

7. You perform the following reaction:
$2A + 2B \leftrightarrow 3C + D$

After waiting three hours, you measure the following concentrations: [A] = 273 mM, [B] = 34.7 mM, [C] = 0.443M, and [D] = 78.9 mM.

a. What is the reaction quotient for the system?

b. If the K_{eq} for this reaction is 1.85×10^2, has the reaction completed?

Solve It

8. Consider the reaction $A + B \leftrightarrow 2C$. The free energy change for this reaction is -25.8 kJ mol^{-1}, and the enthalpy change for the same reaction is 12.3 kJ mol^{-1}.

a. At 273K, what is the entropy change for the reaction? What drives the reaction forward — favorable enthalpy change, favorable entropy change, or both?

b. What is the expression for K_{eq} for the reaction, and what is its value?

c. If a reaction mixture at equilibrium contains 0.744M reactant A and 11.7M product C, what is the concentration of reactant B?

Solve It

Checking Out Factors that Disrupt Equilibrium

After a chemical system has reached equilibrium, that equilibrium can be disrupted, or *perturbed*. Think of systems at equilibrium as people who have finally found their easy chair at the end of a long day. You may rouse them to take out the trash, but they'll return to the easy chair at the first opportunity. This concept is more or less the idea behind *Le Chatelier's Principle:* The equilibrium of a perturbed system shifts in the direction that minimizes the perturbation. Perturbations include changes in concentration, pressure, and temperature.

✔ **Concentration:** If a system is at equilibrium, $Q = K_{eq}$. Adding or removing reactant or product disrupts the equilibrium such that $Q < K_{eq}$ or $Q > K_{eq}$. The equilibrium reasserts itself in response.

   • If reactant is added or product is removed, reactant converts into product.

   • If product is added or reactant is removed, product converts into reactant.

Either way, chemistry occurs until $Q = K_{eq}$ once more. The equilibrium shifts to oppose the perturbation.

✔ **Pressure:** Reactions that include gases as reactants and/or products are particularly sensitive to pressure perturbation. If pressure is suddenly increased, equilibrium shifts toward the side of the reaction that contains fewer moles of gas, thereby decreasing pressure. If pressure is suddenly decreased, equilibrium shifts toward the side of the reaction that contains more moles of gas, thereby increasing pressure. Consider the following reaction:

$$N_2(g) + 3H_2(g) \leftrightarrow 2NH_3(g)$$

A given amount of mass on the reactant side of the equation (as N_2 and H_2) corresponds to double the moles of gas as the same mass on the product side (as NH_3). Imagine that the system is at equilibrium at a low pressure. Now imagine that the pressure suddenly increases, perturbing that equilibrium. Reactant (N_2 and H_2) converts to product (NH_3) so the total moles of gas decrease, thereby lowering the pressure.

If the system suddenly shifts to lower pressure, NH_3 converts to N_2 and H_2, so the total moles of gas increase, thereby raising the pressure. The equilibrium shifts to oppose the perturbation.

✔ **Temperature:** Reactions that absorb or give off heat (that is, most reactions) can be perturbed from equilibrium by changes in temperature. The easiest way to understands this behavior is to explicitly include heat as a reactant or product in the reaction equation:

$$A + B \leftrightarrow C + \text{heat}$$

Imagine that this reaction is at equilibrium. Now imagine that the temperature suddenly increases. Product C absorbs heat, converting to reactants A and B. Because the heat "product" has been decreased, the temperature of the system decreases.

If the temperature suddenly shifts down from equilibrium, reactants A and B convert to product C, releasing heat. The released heat increases the temperature of the system. The equilibrium shifts to oppose the perturbation.

Q. Consider the following reaction, at equilibrium: $2SO_2(g) + O_2(g) \leftrightarrow 2SO_3(g)$

In which direction will each of the following perturbations shift the system?

a. Adding SO_3

b. Removing O_2

c. Pumping unreactive neon gas into the reaction vessel

d. Introducing a metallic catalyst into the system

A. Systems at equilibrium shift to oppose perturbations.

a. Adding SO_3 "pushes" the reaction to the left.

b. Removing O_2 "pulls" the reaction to the left.

c. Pumping unreactive gas into the reaction vessel increases the pressure. The system shifts in the direction that reduces the total moles of gas. This direction is to the right, because the product state contains only 2 moles of gas for every 3 moles of gas in the reactant state.

d. Introducing a catalyst produces no response, because catalysts alter only the kinetics of a chemical reaction, not its equilibrium. (Yes, this part was a trick question!)

---

9. Octane, C_8H_{18}, is a major component of gasoline. Octane vapor combusts to yield carbon dioxide and water vapor:

$$2C_8H_{18}(g) + 25O_2(g) \leftrightarrow 16CO_2(g) + 18H_2O(g) + \text{heat}$$

a. How does increasing the partial pressure of oxygen affect combustion?

b. How does increasing temperature affect combustion?

c. Within engines, fuel and air are compressed within a cylinder and then ignited. The combusting fuel-air mixture expands, driving a piston through the cylinder. How does the expansion affect combustion?

Solve It

10. You conduct several trials of a mysterious, gas phase reaction. The reaction includes an unknown reactant, X, and unknown product, Y: $nX + 2O_2 \rightarrow 4Y$

You observe that the equilibrium yield of Y at 298K is smaller than the yield of Y at 273K. Furthermore, you observe that the yield of Y at 3 atm is smaller than the yield of Y at 1 atm.

a. Does the reaction absorb heat *(endothermic)* or give off heat *(exothermic)?* (See Chapter 15 for more information about these terms.)

b. Which of the following values of n is most plausible: 1, 2, or 3?

Solve It

Answers to Questions on Rate and Equilibrium

You may have raced through these problems, or you may have moved at the speed of a dead snail in winter. But rate is separate from equilibrium — whatever your pace, you've made it this far. Now shift in opposition to any perturbing problems; check your work.

1 The reaction equation makes clear that each mole of methane reactant corresponds to 1 mole of carbon dioxide product and 2 moles of water product. So:

$$d[CH_4] / dt = -d[CO_2] / dt = -0.5d[H_2O] / dt$$

This means that the disappearance of 2.79 mol s^{-1} of methane corresponds to the appearance of 2.79 mol s^{-1} of CO_2 and the appearance of $2 \times 2.79 = 5.58$ mol s^{-1} of H_2O. The concentration of methane changes at half the rate of water, so the concentration of water changes at twice the rate of methane.

2 The observations are consistent with the rate law, Rate = $k[A]^2[B]$. Tripling [A] increased the rate ninefold, and $3^2 = 9$. Doubling [B] doubled the rate, and $2^1 = 2$. The overall reaction order (3) is the sum of the individual reaction orders (the exponents on the individual reactant concentrations), $2 + 1 = 3$.

3 Doubling the concentration of either reactant (D or E) doubles the rate. So, the data are consistent with the rate law, Rate = $k[D][E]$. Solve for k by substituting known values of rate [D] and [E] from any of the trial reactions:

k = Rate / ([D][E]) = $4.8 \times 10^6 M\,s^{-1}$ / $(2.7 \times 10^{-2}M \times 2.7 \times 10^{-2}M)$ = $6.6 \times 10^9 M^{-1}\,s^{-1}$.

Use this calculated value of k to determine the rate in the presence of $1.3 \times 10^{-2}M$ reactant D and $0.92 \times 10^{-2}M$ reactant E:

Rate = $(6.6 \times 10^9 M^{-1}\,s^{-1})(1.3 \times 10^{-2}M)(0.92 \times 10^{-2}M) = 7.9 \times 10^5 M\,s^{-1}$

4 The corning process converts gunpowder from a finely divided powder of undefined particle size into larger particles of defined size. Because the corned particles are larger, combustion occurs more slowly. Corned gunpowder is less likely to explode accidentally than is a fine dust of gunpowder. Because the corned particles are all the same size, combustion occurs at a predictable rate, which is convenient for the user.

5 The change in conditions does result in a faster rate. Rate is limited by activation energy, the difference in energy between reactant (A) and the transition state (A*). If the energy of A is raised and the energy of A* remains constant, then the difference in energy (the activation energy) becomes smaller, and lower activation energies result in faster reactions. The decrease in the energy of product B has no bearing on the rate because it has no effect on the activation energy for the A $\rightarrow$ B reaction — although it would affect the rate of the reverse reaction, B $\rightarrow$ A.

6 Here are the answers regarding equilibrium in the reaction, $N_2(g) + O_2(g) \leftrightarrow 2NO(g)$.

a. The $K_{eq} = (2.63 \times 10^{-7}$ atm$)^2$ / $[(4.76 \times 10^{-2}$ atm$)(9.82 \times 10^{-3}$ atm$)] = 1.48 \times 10^{-10}$.

b. $Q = (2.63 \times 10^{-7}$ atm$)^2$ / $[(4.76 \times 10^{-2}$ atm$)(3.74 \times 10^{-2}$ atm$)] = 3.89 \times 10^{-11}$. Because $Q < K_{eq}$, the reaction would proceed to the right, converting reactant into product.

7   Here are the answers regarding equilibrium in the reaction, $2A + 2B \leftrightarrow 3C + D$.

   a. $Q = [(78.9 \text{ m}M)(443 \text{ m}M)^3] / [(273 \text{ m}M)^2(34.7 \text{ m}M)^2] = 76.4$

   b. If $K_{eq} = 1.85 \times 10^2$, then $Q < K_{eq}$, so the reaction will continue to convert reactant to product.

8   Here are the answers regarding equilibrium in the reaction, $A + B \leftrightarrow 2C$.

   a. $\Delta G = \Delta H - T\Delta S$. Solving for the entropy change, we obtain:

   $\Delta S = (\Delta H - \Delta G) / T = [12.3 \text{ kJ mol}^{-1} - (-25.8 \text{ kJ mol}^{-1})] / 273K = 0.140 \text{ kJ mol}^{-1} \text{ K}^{-1}$. At 273K, the contribution of entropy, $T\Delta S$, is $(0.140 \text{ kJ mol}^{-1} \text{ K}^{-1})(273K) = 38.1 \text{ kJ mol}^{-1}$. This favorable entropy change drives the reaction despite an unfavorable enthalpy change.

   b.

$$K_{eq} = \frac{[C]^2}{[A][B]} = e^{-\Delta G/RT} = e^{\frac{25800 \text{ J mol}^{-1}}{(8.314 \text{ L kPa K}^{-1}\text{mol}^{-1})(273K)}}$$

   c. $[B] = [C]^2 / ([A]K_{eq}) = (11.7M)^2 / [(0.744M)(11.4)] = 16.1M$

9   Here are answers regarding equilibrium in the combustion of octane.

   a. Increasing the partial pressure of oxygen increases the concentration of a reactant, so the system shifts to the right, converting more mass into product.

   b. Increasing temperature is tantamount to adding heat "product," so the system shifts to the left, counteracting the increase in heat content.

   c. As the gases expand within the cylinder, volume increases so pressure decreases (see Chapter 11 for more about this relationship). The decrease in pressure shifts the equilibrium to the right because the product side corresponds to a greater number of moles of gas for a given mass. By increasing total moles of gas, the system shifts to counteract the decrease in pressure.

10   Here are answers regarding equilibrium in the reaction, $nX + 2O_2 \rightarrow 4Y$.

   a. Higher temperatures (more heat) shift the reaction away from product, toward reactant. So, heat must be a product. In other words, the reaction is exothermic.

   b. Increasing pressure leads to lower equilibrium yield. So, the total moles of gas in the product state must exceed the total moles of gas in the reactant state. In other words, $4 > n + 2$. So, the only plausible possibility offered for the value of n is 1.

Chapter 15

Warming Up to Thermochemistry

- -

In This Chapter

▶ Getting a brief overview of thermodynamics

▶ Using heat capacity and calorimetry to measure heat flow

▶ Keeping track of the heats involved in chemical and physical changes

▶ Adding heats together with Hess's Law

- -

*E*nergy shifts between many forms. It may be tricky to detect, but energy is always conserved. Sometimes energy reveals itself as heat. *Thermodynamics* explores how energy moves from one form to another. *Thermochemistry* investigates changes in heat that accompany chemical reactions. In this chapter, we delve into the particulars of thermodynamics and thermochemistry.

Working with the Basics of Thermodynamics

To understand how thermochemistry is done, you need to first understand how the particular form of energy called heat fits into the overall dance of energy and matter.

To study energy, it helps to divide the universe into two parts, the *system* and the *surroundings.* For chemists, the system may consist of the contents of a beaker or tube. This is an example of a *closed system,* one which allows exchange of energy with the surroundings, but doesn't allow exchange of matter. Closed systems are common in chemistry. *Though energy may move between system and surroundings, the total energy of the universe is constant.*

Energy itself is divided into potential energy *(PE)* and kinetic energy *(KE).*

▶ *Potential energy* is energy due to position. *Chemical energy* is a kind of potential energy, arising from the positions of particles within systems.

▶ *Kinetic energy* is the energy of motion. *Thermal energy* is a kind of kinetic energy, arising from the movement of particles within systems.

The total *internal energy* of a system *(E)* is the sum of its potential and kinetic energies. When a system moves between two states (as it does in a chemical reaction), the internal energy

REMEMBER

may change as the system exchanges energy with the surroundings. The difference in energy (ΔE) between the initial and final states derives from heat (q) added to or lost from the system, and from work (w) done by the system or on the system.

We can summarize these energy explanations with the help of a couple of handy formulas:

$$E_{\text{total}} = KE + PE$$

$$\Delta E = E_{\text{final}} - E_{\text{initial}} = q + w$$

What kind of "work" can atoms and molecules do in a chemical reaction? One kind that is easy to understand is *pressure-volume work*. Consider the following reaction:

$$CaCO_3(s) \rightarrow CaO(s) + CO_2(g)$$

Solid calcium carbonate decomposes into solid calcium oxide and carbon dioxide gas. At constant pressure (P), this reaction proceeds with a change in volume (V). The added volume comes from the production of carbon dioxide gas. As gas is made, it expands, pushing against the surroundings. The carbon dioxide gas molecules do work as they push into a greater volume:

$$w = -P\Delta V$$

The negative sign in this equation means that the system loses internal energy because of the work it does on the surroundings. If the surroundings did work on the system, thereby decreasing the system's volume, then the system would gain internal energy.

So, pressure-volume work can partly account for changes in internal energy during a reaction. When pressure-volume work is the only kind of work involved, any remaining changes come from heat flow. *Enthalpy (H)* corresponds to the heat content of a closed system at constant pressure. An *enthalpy change (ΔH)* in such a system corresponds to heat flow. The enthalpy change equals the change in internal energy minus the energy used to perform pressure-volume work:

$$\Delta H = \Delta E - (-P\Delta V) = \Delta E + P\Delta V$$

Like E, P, and V, H is a *state function*, meaning that its value has only to do with the state of the system, and nothing to do with how the system got to that state. Heat (q) is *not* a state function, but is simply a form of energy that flows from warmer objects to cooler objects.

TIP

Now breathe. The practical consequences of all this theory are the following:

- ✔ Chemical reactions usually involve the flow of energy in the form of heat, q.

- ✔ Chemists monitor changes in heat by measuring changes in temperature.

- ✔ At constant pressure, the change in heat content equals the change in enthalpy, ΔH.

- ✔ Knowing ΔH values helps to explain and predict chemical behavior.

Q. Gas is heated within a sealed cylinder. The heat causes the gas to expand, pushing a movable piston outward to increase the volume of the cylinder to 4.63L. The initial and final pressures of the system are both 1.15 atm. The gas does 304J of work on the piston. What was the initial volume of the cylinder? (*Note:* 101.3J = 1 L·atm)

A. **2.02L.** You're given an amount of work, a constant pressure, and the knowledge that the volume of the system changes. So, the equation to use here is $w = -P\Delta V = -P(V_{final} - V_{initial})$. Because the system does work on the piston (and not the other way around), the sign of w is negative. Substituting your known values into the equation gives you:

$$-304J = -1.15 \text{ atm} (4.63L - V_{initial})$$

Converting joules to liter-atmospheres yields:

$$-3.00 \text{ L·atm} = -1.15 \text{ atm} (4.63L - V_{initial})$$

Solving for $V_{initial}$ gives you:

$$V_{initial} = (5.32 \text{ L·atm} - 3.00 \text{ L·atm}) / 1.15 \text{ atm} = 2.02L$$

---

1. A fuel combusts at 3.00 atm constant pressure. The reaction releases 75.0 kJ of heat and causes the system to expand from 7.50L to 20.0L. What is the change in internal energy? (*Note:* 101.3J = 1 L·atm)

Solve It

Holding Heat: Heat Capacity and Calorimetry

Heat is a form of energy that flows from warmer objects to cooler objects. But how much heat can an object hold? If objects have the same heat content, does that mean they're the same temperature? You can measure different temperatures, but how do these temperatures relate to heat flow? These kinds of questions revolve around the concept of *heat capacity,* the amount of heat required to raise the temperature of a system by 1K.

It takes longer to boil a large pot of water than a small pot of water. With the burner set on high, the same amount of heat flows into each pot, but the larger pot of water has a higher heat capacity. So, it takes more heat to increase the temperature of the larger pot.

You'll encounter heat capacity in different forms, each of which is useful in different scenarios. Any system has a heat capacity. But how can you best compare heat capacities between chemical systems? You use *molar heat capacity* or *specific heat capacity* (or just *specific heat*). Molar heat capacity is simply the heat capacity of 1 mole of a substance. Specific heat capacity is simply the heat capacity of 1 gram of a substance. How do you know whether you're dealing with heat capacity, molar heat capacity, or specific heat capacity? Look at the units.

Heat capacity: Energy / K

Molar heat capacity: Energy / (mol·K)

Specific heat capacity: Energy / (g·K)

Fine, but what are the units of energy? Well, that depends. The SI unit of energy is the *joule (J)* (see Chapter 2 for more about the International System of units), but the units of *calorie (cal)* and *liter-atmosphere (L·atm)* are also used. Here's how the joule, the calorie, and the liter-atmosphere are related:

- ✔ 1J = 0.2390 cal
- ✔ 101.3J = 1 L·atm

Note that in everyday language, a "calorie" actually refers to a "Calorie" or a kilocalorie — a calorie of cheesecake is 1,000 times larger than you think it is.

Calorimetry is a family of techniques that puts all this thermochemical theory to use. When chemists do calorimetry, they initiate a reaction within a defined system, and then measure any temperature change that occurs as the reaction progresses. There are a few variations on this theme.

- ✔ *Constant-pressure calorimetry* directly measures an enthalpy change (ΔH) for a reaction because it monitors heat flow at constant pressure: $\Delta H = q_P$

  Typically, heat flow is observed through changes in the temperature of a reaction solution. If a reaction warms a solution, then that reaction must have released heat into the solution. In other words, the change in heat content of the reaction ($q_{reaction}$) has the same magnitude as the change in heat content for the solution ($q_{solution}$) but has opposite sign: $q_{solution} = -q_{reaction}$

  So, measuring $q_{solution}$ allows you to calculate $q_{reaction}$, but how can you measure $q_{solution}$? You do so by measuring the difference in temperature (ΔT) before and after the reaction:

  $q_{solution}$ = (mass of solution) × (specific heat of solution) × ΔT

  In other words, $q = mC_P\Delta T$.

  Here, "m" is the mass of the solution and C_P is the specific heat capacity of the solution at constant pressure. ΔT is equal to $T_{final} - T_{initial}$.

  When you use this equation, be sure that all your units match. For example, if your C_P has units of $J\ g^{-1}\ K^{-1}$, don't expect to calculate heat flow in kilocalories.

- ✔ *Constant-volume calorimetry* directly measures a change in internal energy (ΔE, not ΔH) for a reaction because it monitors heat flow at constant volume. Often, ΔE and ΔH are very similar values.

  A common variety of constant-volume calorimetry is *bomb calorimetry,* a technique in which a reaction (often, a combustion reaction) is triggered within a sealed vessel called a bomb. The vessel is immersed in a water bath of known volume. The temperature of the water is measured before and after the reaction. Because the heat capacity of the water and the calorimeter are known, you can calculate heat flow from the change in temperature.

Q. Paraffin wax is sometimes incorporated within sheetrock to act as an insulator. During the day, the wax absorbs heat and melts. During the cool nights, the wax releases heat and solidifies. At sunrise, a small hunk of solid paraffin within a wall has a temperature of 298K. The rising sun warms the wax, which has a melting temperature of 354K. If the hunk of wax has 0.257g mass and a specific heat capacity of 2.50J g^{-1} K^{-1}, how much heat must the wax absorb to bring it to its melting point?

A. **36.0 joules.** You're given two temperatures, a mass, and a specific heat capacity. You're asked to find an amount of heat energy. You've got all you need to proceed with $q = mC_P \Delta T$.

$q = (0.257g)(2.50J\ g^{-1}\ K^{-1})(354K - 298K) = 36.0J$

2. A 375g plug of lead is heated and placed into an insulated container filled with 0.500L water. Prior to the immersion of the lead, the water is at 293K. After a time, the lead and the water assume the same temperature, 297K. The specific heat capacity of lead is 0.127J g^{-1} K^{-1}, and the specific heat capacity of water is 4.18J g^{-1} K^{-1}. How hot was the lead before it entered the water? (**Hint:** You'll need to use the density of water.)

Solve It

3. At some point, all laboratory chemists learn the same hard lesson: Hot glass looks just like cold glass. Heath discovered this when he picked up a hot beaker someone left on his bench. At the moment Heath grasped the 413K glass beaker, 567J of heat flowed out of the beaker and into his hand. The glass of the beaker has a heat capacity of 0.84J g^{-1} K^{-1}. If the beaker was 410K the instant Heath dropped it, then what is the collective mass of the shards of beaker now littering the lab floor?

Solve It

4. 25.4g of NaOH are dissolved in water within an insulated calorimeter. The heat capacity of the resulting solution is 4.18×10^3 J K^{-1}. The temperature of the water prior to the addition of NaOH was 296K. If NaOH releases 44.2 kJ mol^{-1} as it dissolves, what is the final temperature of the solution?

Solve It

Absorbing and Releasing Heat: Endothermic and Exothermic Reactions

So, as you find out in the previous section, you can monitor heat flow by measuring changes in temperature. But what does any of this have to do with chemistry? Chemical reactions transform both matter and energy. Though reaction equations usually list only the matter components of a reaction, you can consider heat energy as a reactant or product as well. When chemists are interested in heat flow during a reaction (and when the reaction is run at constant pressure), they may list an enthalpy change *(ΔH)* to the right of the reaction equation. As we explain in the previous section, at constant pressure, heat flow equals *ΔH*.

$$q_P = \Delta H = H_{final} - H_{initial}$$

If the *ΔH* listed for a reaction is negative, then that reaction releases heat as it proceeds — the reaction is *exothermic.* If the *ΔH* listed for the reaction is positive, then that reaction absorbs heat as it proceeds — the reaction is *endothermic.* In other words, exothermic reactions release heat as a product, and endothermic reactions consume heat as a reactant.

The sign of the *ΔH* tells us the direction of heat flow, but what about the magnitude? The coefficients of a chemical reaction represent molar equivalents (see Chapter 8 for details). So, the value listed for the *ΔH* refers to the enthalpy change for one molar equivalent of the reaction. Here's an example:

$$CH_4(g) + 2O_2(g) \rightarrow CO_2(g) + 2H_2O(g) \qquad \Delta H = -802 \text{ kJ}$$

The reaction equation shown describes the combustion of methane, a reaction you might expect to release heat. The enthalpy change listed for the reaction confirms this expectation: For each mole of methane that combusts, 802 kJ of heat are released. The reaction is highly exothermic. Based on the stoichiometry of the equation, you can also say that 802 kJ of heat are released for every 2 moles of water produced. (Flip to Chapter 9 for the scoop on stoichiometry.)

So, reaction enthalpy changes (or reaction "heats") are a useful way to measure or predict chemical change. But they're just as useful in dealing with physical changes, like freezing and melting, evaporating and condensing, and others. For example, water (like most substances) absorbs heat as it melts (or *fuses*) and as it evaporates:

> Molar enthalpy of fusion: $\Delta H_{fus} = 6.01$ kJ
>
> Molar enthalpy of vaporization: $\Delta H_{vap} = 40.68$ kJ

The same sorts of rules apply to enthalpy changes listed for chemical changes and physical changes. Here's a summary of the rules that apply to both:

✔ **The heat absorbed or released by a process is proportional to the moles of substance that undergo that process.** Two moles of combusting methane release twice as much heat as does 1 mole of combusting methane.

✔ **Running a process in reverse produces heat flow of the same magnitude but of opposite sign as running the forward process.** Freezing 1 mole of water releases the same amount of heat that is absorbed when 1 mole of water melts.

Q. Here is a balanced chemical equation for the oxidation of hydrogen gas to form liquid water, along with the corresponding enthalpy change:

$$2H_2(g) + O_2(g) \rightarrow 2H_2O(l) \quad \Delta H = -572 \text{ kJ}$$

How much electrical energy must be expended to perform electrolysis of 3.76 mol of liquid water, converting that water into hydrogen gas and oxygen gas?

A. **1.08×10^3 kJ.** First, recognize that the given enthalpy change is for the reverse of the electrolysis reaction, so you must reverse its sign from –572 to 572. Second, recall that heats of reaction are proportional to the amount of substance reacting (2 moles of H_2O in this case), so the calculation is:

$$3.76 \text{ mol } H_2O \times (572 \text{ kJ} / 2 \text{ mol } H_2O) = 1.08 \times 10^3 \text{ kJ}$$

---

5. Carbon dioxide gas can be decomposed into oxygen gas and carbon monoxide:

$$2CO_2(g) \rightarrow O_2(g) + 2CO(g) \quad \Delta H = 486 \text{ kJ}$$

How much heat is released or absorbed when 9.67g of carbon monoxide combine with oxygen to form carbon dioxide?

Solve It

6. How much heat must be added to convert 4.77 mol of 268K ice into steam? The specific heat capacity of ice is 38.1J mol^{-1} K^{-1}. The specific heat capacity of water is 75.3J mol^{-1} K^{-1}. The molar enthalpies of fusion and vaporization are 6.01 kJ and 40.68 kJ, respectively.

Summing Heats with Hess's Law

For the chemist, *Hess's Law* is a valuable tool for dissecting heat flow in complicated, multi-step reactions. For the confused or disgruntled chemistry student, Hess's Law is a breath of fresh air. In essence, the law confirms that heat behaves the way we'd like it to behave: predictably.

Imagine that the product of one reaction serves as the reactant for another reaction. Now imagine that the product of the second reaction serves as the reactant for a third reaction. What you have is a set of coupled reactions, connected in series like the cars of a train:

$$A \rightarrow B \quad \text{and} \quad B \rightarrow C \quad \text{and} \quad C \rightarrow D$$

So,

$$A \rightarrow B \rightarrow C \rightarrow D$$

You can think of these three reactions adding up to one big reaction, $A \rightarrow D$. What is the overall enthalpy change associated with this reaction ($\Delta H_{A \rightarrow D}$)? Here's the good news:

$$\Delta H_{A \rightarrow D} = \Delta H_{A \rightarrow B} + \Delta H_{B \rightarrow C} + \Delta H_{C \rightarrow D}$$

Enthalpy changes are additive. But the good news gets even better. Imagine that you're trying to figure out the total enthalpy change for the following multistep reaction:

$$X \rightarrow Y \rightarrow Z$$

Here's a wrinkle: For technical reasons, you can't measure this enthalpy change ($\Delta H_{X \rightarrow Z}$) directly, but must calculate it from tabulated values for $\Delta H_{X \rightarrow Y}$ and $\Delta H_{Y \rightarrow Z}$. No problem, right? You simply look up the tabulated values and add them. But here's another wrinkle: when you look up the tabulated values, you find the following:

$$\Delta H_{X \rightarrow Y} = -37.5 \text{ kJ mol}^{-1}$$

$$\Delta H_{Z \rightarrow Y} = -10.2 \text{ kJ mol}^{-1}$$

Gasp! You need $\Delta H_{Y \rightarrow Z}$, but you're provided only $\Delta H_{Z \rightarrow Y}$! Relax. As we note in the previous section, the enthalpy change for a reaction has the same magnitude and opposite sign as the reverse reaction. So, if $\Delta H_{Z \rightarrow Y} = -10.2 \text{ kJ mol}^{-1}$, then $\Delta H_{Y \rightarrow Z} = 10.2 \text{ kJ mol}^{-1}$. It really is that simple. So,

$$\Delta H_{X \rightarrow Z} = \Delta H_{X \rightarrow Y} + (-\Delta H_{Z \rightarrow Y}) = -37.5 \text{ kJ mol}^{-1} + 10.2 \text{ kJ mol}^{-1} = -27.3 \text{ kJ mol}^{-1}$$

Thanks be to Hess.

EXAMPLE

$Q.$ Calculate the reaction enthalpy for the following reaction:

$$PCl_5(g) \rightarrow PCl_3(g) + Cl_2(g)$$

Use the following data:

Reaction 1:	$4PCl_3(g) \rightarrow P_4(s) + 6Cl_2(g)$	$\Delta H = 821 \text{ kJ}$
Reaction 2:	$P_4(s) + 10Cl_2(g) \rightarrow 4PCl_5(g)$	$\Delta H = -1156 \text{ kJ}$

$A.$ **83.8 kJ.** Reaction enthalpies are given for two reactions. Your task is to manipulate and add Reactions 1 and 2 so the sum is equivalent to the target reaction. First, reverse Reactions 1 and 2 to obtain Reactions 1' and 2', and add the two reactions:

Reaction 1':	$P_4(s) + 6Cl_2(g) \rightarrow 4PCl_3(g)$	$\Delta H = -821 \text{ kJ}$
Reaction 2':	$4PCl_5(g) \rightarrow P_4(s) + 10Cl_2(g)$	$\Delta H = 1156 \text{ kJ}$

Sum:	$4PCl_5(g) \rightarrow 4PCl_3(g) + 4Cl_2(g)$	$\Delta H = 335 \text{ kJ}$

Identical species that appear on opposite sides of the equations cancel out (as occurs with species P_4 and Cl_2). Finally, divide the sum by 4 to yield the target reaction equation:

$$PCl_5(g) \rightarrow PCl_3(g) + Cl_2(g) \quad \Delta H = 83.8 \text{ kJ}$$

---

7. Calculate the reaction enthalpy for the following reaction:

$$N_2O_4(g) \rightarrow N_2(g) + 2O_2(g)$$

Use the following data:

Reaction 1:	$N_2(g) + 2O_2(g) \rightarrow 2NO_2(g)$	$\Delta H = 1032 \text{ kJ}$
Reaction 2:	$2NO_2(g) \rightarrow N_2O_4(g)$	$\Delta H = -886 \text{ kJ}$

Solve It

8. Calculate the reaction enthalpy for the following reaction:

$C(s) + H_2O(g) \rightarrow CO(g) + H_2(g)$

Use the following data:

Reaction 1:	$C(s) + O_2(g) \rightarrow CO_2(g)$	$\Delta H = -605$ kJ
Reaction 2:	$2CO(g) + O_2(g) \rightarrow 2CO_2(g)$	$\Delta H = -966$ kJ
Reaction 3	$2H_2(g) + O_2(g) \rightarrow 2H_2O(g)$	$\Delta H = -638$ kJ

Answers to Questions on Thermochemistry

Check your answers to the practice problems to see whether you've truly felt the heat.

1 **78.8 kJ decrease.** You're given a constant pressure, an amount of heat, a change in volume, and the knowledge that there's been an change in internal energy. Changes in heat content at constant pressure are equivalent to a change in enthalpy. So, the equation to use here is:

$$\Delta H = \Delta E + P\Delta V$$

To do the math properly, you must make sure that all of your units match, so temporarily convert the given heat energy from kJ to L·atm: 75.0 kJ = 7.50×10^4J = 7.40×10^2 L·atm.

Because heat is released from the system, the change in enthalpy (ΔH) is negative. Substitute your known values into the equation:

-7.40×10^2 L·atm $= \Delta E + 3.00$ atm (20.0L – 7.50L)

Solving for ΔE gives you –778 L·atm, which is equivalent to –78.8 kJ. Because the sign is negative, the internal energy of the system decreases by 78.8 kJ.

2 **473K.** The key to setting up this problem is to realize that whatever heat flows out of the lead flows into the water, so: $q_{lead} = -q_{water}$. Calculate each quantity of heat by using $q = mC_P\Delta T$. Recall that $\Delta T = T_{final} - T_{initial}$. The unknown in the problem is the initial temperature of lead.

$q_{lead} = (375g)(0.127$J g^{-1} K$^{-1})(297K - T_{initial})$

To calculate q_{water}, you must first calculate the mass of 0.500L water by using the density of water: 0.500L × (1.00 kg L^{-1}) = 0.500 kg = 500g

$q_{water} = (500g)(4.18$J g^{-1} K$^{-1})(297K - 293K) = 8.36 \times 10^3$J

Setting q_{lead} equal to $-q_{water}$ and solving for $T_{initial}$ yields 473K:

$(375g)(0.127$J g^{-1} K$^{-1})(297K - T_{initial}) = -8.36 \times 10^3$J

$297K - T_{initial} = -176K$

$T_{initial} = 473K$

3 **2.3×10^2g.** To solve this problem, apply $q = mC_P\Delta T$ to the beaker. The unknown is the mass of the beaker, m. Because heat flowed out of the beaker, the sign of q must be negative.

-576J $= (m)(0.84$J g^{-1} K$^{-1})(410K - 413K)$

Solving for m yields 2.3×10^2g.

4 **303K.** Solve this problem in two parts. First, calculate the amount of heat released during dissolution of the NaOH:

$$\frac{25.4\text{g NaOH}}{1} \times \frac{\text{mol NaOH}}{40.0\text{g NaOH}} \times \frac{44.2\text{ kJ}}{\text{mol NaOH}} = 28.1 \text{ kJ}$$

Because you're given the solution's heat capacity — not its molar heat capacity or its specific heat capacity — you can simply substitute the released heat, q, into the equation

$q = $ (heat capacity)$(T_{final} - T_{initial})$

Because the given heat capacity uses units of joules, you must convert the heat into joules from kilojoules before plugging in the value:

2.81×10^4J $= (4.18 \times 10^3$ J K$^{-1})(T_{final} - 296K)$

Solving for T_{final} yields 303K.

5 **–83.9 kJ are released.** Solve this problem with a chain of conversion factors. Convert from grams of CO to moles, and then from moles to kJ. Be sure to adjust the sign of the enthalpy and incorporate the stoichiometry of the given reaction equation:

$$\frac{9.67\text{g CO}}{1} \times \frac{\text{mol CO}}{28.0\text{g CO}} \times \frac{-486 \text{ kJ}}{2 \text{ mol CO}} = -83.9 \text{ kJ}$$

Because the sign is negative, 83.9 kJ are released (not absorbed).

6 2.60×10^5**J (or** 2.60×10^2 **kJ).** The total heat required is the sum of several individual heats: heat to warm the ice to the melting point (273K), heat to convert the ice to liquid water, heat to warm the liquid water to the boiling point, and heat to convert the liquid water to steam. Be careful to match your units (J versus kJ):

q (warm ice to melting point) = $(4.77 \text{ mol})(38.1\text{J mol}^{-1}\text{ K}^{-1})(273\text{K} - 268\text{K}) = 909\text{J}$

q (convert ice to liquid water) = $(4.77 \text{ mol})(6.01 \times 10^3\text{J mol}^{-1}) = 2.87 \times 10^4\text{J}$

q (warm water to boiling point) = $(4.77 \text{ mol})(75.3\text{J mol}^{-1}\text{ K}^{-1})(373\text{K} - 273\text{K}) = 3.59 \times 10^4\text{J}$

q (convert liquid water to steam) = $(4.77 \text{ mol})(4.068 \times 10^4\text{J mol}^{-1}) = 1.94 \times 10^5\text{J}$

The sum of all heats is 2.60×10^5J (or 2.60×10^2 kJ).

7 **–146 kJ.** Reverse Reaction 1 to get Reaction 1' (–1032kJ). Reverse Reaction 2 to get Reaction 2' (886kJ). Add Reactions 1' and 2', yielding ΔH = –146 kJ.

8 **197 kJ.** Reverse Reaction 2 and divide it by 2 to get Reaction 2' (483kJ). Reverse Reaction 3 and divide it by 2 to get Reaction 3' (319kJ). Add Reactions 1, 2', and 3', yielding ΔH = 197 kJ. You need to reverse Reactions 2 and 3 to put the right compounds on the reactant and product sides of the equation. Reaction 1 already has the right orientation. Reactions 2 and 3 must be divided by 2 so the stoichiometry of the final equation matches the stoichiometry of the target equation (namely, 1 mole each of H_2O, CO, and H_2). Even though this division temporarily creates noninteger coefficients for O_2 in these equations, the noninteger coefficients cancel out when you add reactions 1, 2', and 3'.

Part IV
Swapping Charges

The 5th Wave
By Rich Tennant

In this part . . .

Atoms are built with three kinds of particles, two of which are charged. This means that charge is important in chemistry. Swapping or transferring charge between reactants alters their properties. Two critical classes of reactions revolve around such movements of charge: acid-base reactions and oxidation-reduction reactions. In this part, we give you the tools to deal with these charge-centric reactions. In addition, we describe nuclear chemistry, which deals with transformations in atomic nuclei, the positively charged hearts of atoms.

Chapter 16

Giving Acids and Bases the Litmus Test

● ●

In This Chapter
▶ Checking out the many definitions of acid and base
▶ Calculating acidity and basicity
▶ Determining the strength of acids and bases with the help of dissociation

● ●

*1*f you've read any comic books, watched any superhero flicks, or even tuned in to one of those fictional solve-the-crime-in-50-minutes shows on TV, you've likely come across a reference to acids being dangerous substances. Acids are generally thought of as something which evil villains intend to spray in the face of a hero or heroine, but somehow usually manage to spill on themselves. However, you encounter and even ingest a wide variety of fairly innocuous acids in everyday life. Citric acid, present in citrus fruits such as lemons and oranges, is very ingestible, as is acetic acid, also known as vinegar.

Strong acids can indeed burn the skin and must be handled with care in the laboratory. However, strong bases can burn skin as well. Chemists must have a more sophisticated understanding of the differences between an acid and a base and their relative strengths than simply their propensity to burn. This chapter focuses on how you can identify acids and bases, as well as several ways to determine their strengths.

Three Complementary Methods for Defining Acids and Bases

As chemists came to understand acids and bases as more than just "stuff that burns," their understanding of how to define them evolved as well. It's often said that acids taste sour, while bases taste bitter, but we do *not* recommend that you go around tasting chemicals in the laboratory to identify them as acids or bases. In the following sections, we explain three much safer methods you can use to tell the difference between the two.

Method 1: Arrhenius sticks to the basics

Svante Arrhenius was a Swedish chemist who is credited not only with the acid-base determination method that is named after him, but with an even more fundamental chemical concept: that of *dissociation.* Arrhenius proposed an explanation in his PhD thesis for a phenomenon that, at the time, had chemists all over the world scratching their heads. What had them perplexed was this: Although neither pure salt nor pure water is a good conductor of electricity, solutions in which salts are dissolved in water tend to be excellent conductors of electricity.

Arrhenius proposed that aqueous solutions of salts conducted electricity because the bonds between atoms in the salts had been broken simply by mixing them into the water, forming ions. Although the ion had been defined several decades earlier by Michael Faraday, chemists generally believed at the time that ions could only form through *electrolysis,* or the breaking of chemical bonds using electric currents, so Arrhenius's theory was met with some skepticism. Ironically, although his thesis committee wasn't overly impressed and gave him a grade just barely sufficient to pass, Arrhenius eventually managed to win over the scientific community with his research. He was awarded the Nobel Prize in Chemistry in 1903 for the same ideas that nearly cost him his doctorate.

Arrhenius subsequently expanded his theories to form one of the most widely used and straightforward definitions of acids and bases. Arrhenius said that acids are substances that form hydrogen (H^+) ions when they dissociate in water, while bases are substances that form hydroxide (OH^-) ions when they dissociate in water.

Peruse Table 16-1 for a list of common acids and bases and note that all of the acids in the list contain a hydrogen and most of the bases contain a hydroxide. The Arrhenius definition of acids and bases is straightforward and works for many common acids and bases, but it's limited by its narrow definition of bases.

Table 16-1	Common Acids and Bases		
Acid Name	*Chemical Formula*	*Base Name*	*Chemical Formula*
Acetic Acid	CH_3COOH	Ammonia	NH_3
Citric Acid	$C_5H_7O_5COOH$	Calcium Hydroxide	$Ca(OH)_2$
Hydrochloric Acid	HCl	Magnesium Hydroxide	$Mg(OH)_2$
Hydrofluoric Acid	HF	Potassium Hydroxide	KOH
Nitric Acid	HNO_3	Sodium Carbonate	Na_2CO_3
Nitrous Acid	HNO_2	Sodium Hydroxide	$NaOH$
Sulfuric Acid	H_2SO_4		

Method 2: Brønsted-Lowry tackles bases without a hydroxide ion

You no doubt noticed that some of the bases in Table 16-1 don't contain a hydroxide ion, which means that the Arrhenius definition of acids and bases can't apply. When chemists realized that several substances behaved like bases but didn't contain a hydroxide ion, they reluctantly acknowledged that another determination method was needed. Independently proposed by Johannes Brønsted and Thomas Lowry in 1923 and therefore named after both of them, the Brønsted-Lowry method for determining acids and bases accounts for those pesky non-hydroxide–containing bases.

Under the Brønsted-Lowry definition, an acid is a substance that donates a hydrogen ion (H⁺) in an acid-base reaction, while a base is a substance that accepts that hydrogen ion from the acid. When ionized to form a hydrogen cation (an atom with a positive charge), hydrogen loses its one and only electron and is left with only a single proton. For this reason, Brønsted-Lowry acids are often called *proton donors* and Brønsted-Lowry bases are called *proton acceptors.*

The best way to spot Brønsted-Lowry acids and bases is to keep careful track of hydrogen ions in a chemical equation. Consider, for example, the dissociation of the base sodium carbonate in water. Note that although sodium carbonate is a base, it doesn't contain a hydroxide ion.

$$Na_2CO_3 + 2H_2O \rightarrow H_2CO_3 + 2NaOH$$

This is a simple double replacement reaction (see Chapter 8 for an introduction to these types of reactions). A hydrogen ion from water switches places with the sodium of sodium carbonate to form the products carbonic acid and sodium hydroxide. By the Brønsted-Lowry definition, water is the acid because it gave up its hydrogen to Na_2CO_3. This makes Na_2CO_3 the base because it accepted the hydrogen ion from H_2O.

What about the substances on the right-hand side of the equation? Brønsted-Lowry theory calls the products of an acid-base reaction the *conjugate acid* and *conjugate base.* The conjugate acid is produced when the base accepts a proton (in this case, H_2CO_3), while the conjugate base is formed when the acid loses its hydrogen (in this case, NaOH). This reaction also brings up a very important point about the strength of each of these acids and bases. Although sodium carbonate is a very strong base, its conjugate acid, carbonic acid, is a very weak acid. Similarly, water is an extremely weak acid, and its conjugate base, sodium hydroxide, is a very strong base. Weak acids always form strong conjugate bases and vice versa. The same is true of strong acids.

Method 3: Lewis relies on electron pairs

In the same year that Brønsted and Lowry proposed their definition of acids and bases, an American chemist named Gilbert Lewis proposed an alternate definition that not only encompassed Brønsted-Lowry theory, but also accounted for acid-base reactions in which a hydrogen ion isn't exchanged. Lewis's definition relies on tracking lone pairs of electrons. Under his theory, a base is any substance that donates a pair of electrons to form a coordinate covalent bond with another substance, while an acid is a substance that accepts that electron pair in such a reaction. As we explain in Chapter 5, a coordinate covalent bond is a covalent bond in which both of the bonding electrons are donated by one of the atoms forming the bond.

All Brønsted-Lowry acids are Lewis acids, but in practice, the term Lewis acid is generally reserved for Lewis acids that don't also fit the Brønsted-Lowry definition. The best way to spot a Lewis acid-base pair is to draw a Lewis dot structure of the reacting substances, noting the presence of lone pairs of electrons. (We introduce Lewis structures in Chapter 5.) For example, consider the reaction between ammonia (NH_3) and boron trifluoride (BF_3):

$$NH_3 + BF_3 \rightarrow NH_3BF_3$$

At first glance, neither the reactants nor the product appear to be acids or bases, but the reactants are revealed as a Lewis acid-base pair when drawn as Lewis dot structures as in Figure 16-1. Ammonia donates its lone pair of electrons to the bond with boron trifluoride, making ammonia the Lewis base and boron trifluoride the Lewis acid.

Figure 16-1:
The Lewis dot structures of ammonia and boron trifluoride.

Sometimes you can identify the Lewis acid and base in a compound without drawing the Lewis dot structure. You can do this by identifying reactants that are electron rich (bases) or electron poor (acids). Metal cations, for example, are electron poor and tend to act as a Lewis acid in a reaction, accepting a pair of electrons.

TIP

In practice, it's much simpler to use the Arrhenius or Brønsted-Lowry definitions of acid and base, but you'll need to use the Lewis definition when hydrogen ions aren't being exchanged. You can pick and choose among the definitions when you're asked to identify the acid and base in a reaction.

EXAMPLE

Q. Identify the acid and base in the following reaction and label their conjugates.

$NH_3 + H_2O \rightarrow NH_4^+ + OH^-$

A. This is a classic Brønsted-Lowry acid-base pair. Water loses a proton to ammonia, forming a hydroxide anion (an atom with a negative charge). This makes water the proton donor, or Brønsted-Lowry acid, and OH⁻ its conjugate base. Ammonia accepts the proton from water, forming ammonium (NH_4^+). This makes ammonia the proton acceptor, or Brønsted-Lowry base, and NH_4^+ its conjugate acid.

1. Consider the following reaction. Label the acid, base, conjugate acid, and conjugate base, and comment on their strengths. How can water act as an acid in one reaction and a base in another?

$HCl + H_2O \rightarrow H_3O^+ + Cl^-$

2. Use the Arrhenius definition to identify the acid or base in each reaction and explain how you know.

a. $NaOH(s) + H_2O \rightarrow Na^+(aq) + OH^-(aq) + H_2O$

b. $HF(g) + H_2O \rightarrow H^+(aq) + F^-(aq) + H_2O$

Solve It

3. Identify the Lewis acid and base in each reaction below. Draw Lewis dot structures for the first two, and determine the Lewis acid and base in the third reaction without a dot structure.

a. $6H_2O + Cr^{3+} \rightarrow Cr(OH_2)_6^{3+}$

b. $2NH_3 + Ag^+ \rightarrow Ag(NH_3)_2^+$

c. $2Cl^- + HgCl_2 \rightarrow HgCl_4^{2-}$

Solve It

Measuring Acidity and Basicity: pH, pOH, and K_W

A substance's identity as an acid or a base is only one of many things that a chemist may need to know about it. Sulfuric acid and water, for example, can both act as acids, but using sulfuric acid to wash your face in the morning would be a grave error indeed. Sulfuric acid and water differ greatly in *acidity,* a measurement of an acid's strength. A similar quantity, called *basicity,* measures a base's strength.

Acidity and basicity are measured in terms of quantities called *pH* and *pOH,* respectively. Both are simple scales ranging from 0 to 14, with low numbers on the pH scale representing a higher acidity and therefore a stronger acid. On both scales, a measurement of 7 indicates a *neutral solution.* On the pH scale, any number lower than 7 indicates that the solution is acidic, with acidity increasing as pH decreases, while any number higher than 7 indicates a basic solution, with basicity increasing as pH increases. In other words, the farther the pH gets away from 7, the more acidic or basic a substance gets. pOH shows exactly the same relationship between distance from 7 and acidity or basicity, only this time low numbers indicate very basic solutions, while high numbers indicate very acidic solutions.

pH is calculated using the formula pH = –log[H⁺], where the brackets around H⁺ indicate that it's a measurement of the concentration of hydrogen ions in moles per liter (or molarity; see Chapter 12). pOH is calculated using a similar formula, with OH⁻ concentration replacing the H⁺ concentration: pOH = –log[OH⁻]. (The word *log* in each formula stands for *logarithm*.)

Because a substance with high acidity must have low basicity, a low pH indicates a high pOH for a substance and vice versa. In fact, a very convenient relationship between pH and pOH allows you to solve for one when you have the other: pH + pOH = 14.

You'll often be given a pH or pOH and be asked to solve for the H⁺ or OH⁻ concentrations instead of the other way around. The logarithms in the pH and pOH equations make it tricky to solve for [H⁺] or [OH⁻], but if you remember that a log is undone by raising 10 to both sides of an equation, you quickly arrive at a convenient formula for [H⁺], namely [H⁺] = 10^{-pH}. Similarly, [OH⁻] can be calculated using the formula [OH⁻] = 10^{-pOH}. As with pH and pOH, a convenient relationship exists between [H⁺] and [OH⁻], which multiply together to equal a constant. This constant, called the *ion product constant for water,* or K_W, is calculated as follows:

$$K_W = [H^+] \times [OH^-] = 1 \times 10^{-14}$$

Q. Calculate the pH and pOH of a solution with an [H⁺] of 1×10^{-8}. Is the solution acidic or basic? Do the same for a solution with an [OH⁻] of 2.3×10^{-11}.

A. For a solution with an [H⁺] of 1×10^{-8}: **pH = 8; pOH = 6; the solution is a base.** You have been given the H⁺ concentration, so first solve for the pH by plugging [H⁺] into the formula for pH, giving

pH = –log[1×10^{-8}] = 8

Plugging this value into the equation relating pH and pOH gives you a value for pOH.

14 – pH = 14 – 8 = 6 = pOH

A pH of 8 indicates that this solution is very slightly basic. It's not just a coincidence that the exponent of the H⁺ concentration is equal to the pH. This is true whenever the coefficient of the H⁺ concentration is 1.

For a solution with an [OH⁻] of 2.3×10^{-11}: **pOH = 10.6; pH = 3.4; the solution is an acid.** The second portion of the problem gives you an [OH⁻], in which case you can either use the ion-product constant for water to calculate the [H⁺] and then solve for pH and pOH as you did before, or, more simply, you can first calculate the pOH and use it to find the pH. Plugging an [OH⁻] of 2.3×10^{-11} into the pOH equation yields

pOH = –log[2.3×10^{-11}] = 10.6

Plug this value into the relation between pH and pOH to get pH.

14 – pOH = 14 – 10.6 = 3.4 = pH

This low pH indicates that the substance is a relatively strong acid.

4. Determine the pH given the following values:

    a. $[H^+] = 1 \times 10^{-13}$

    b. $[H^+] = 1.58 \times 10^{-9}$

    c. $[OH^-] = 2 \times 10^{-7}$

    d. $[OH^-] = 1 \times 10^{-7}$

Solve It

5. Determine the pOH given the following values:

    a. $[H^+] = 2 \times 10^{-8}$

    b $[OH^-] = 5.1 \times 10^{-11}$

Solve It

6. Determine whether the following are acidic, basic, or neutral:

    a. $[OH^-] = 2.5 \times 10^{-7}$

    b. $[OH^-] = 3.1 \times 10^{-13}$

    c. $[H^+] = 4.21 \times 10^{-5}$

    d. $[H^+] = 8.9 \times 10^{-10}$

Solve It

7. Determine the $[H^+]$ from the following pH or pOH values:

    a. pH = 3.3

    b. pH = 7.69

    c. pOH = 10.21

    d. pOH = 1.26

Solve It

Finding Strength through Dissociation: K_a and K_b

Arrhenius's concept of dissociation (which we cover earlier in this chapter) gives us another convenient way of measuring the strength of an acid or base. Although water tends to dissociate all acids and bases, the degree to which they dissociate depends on their strength. Strong acids such as HCl, HNO_3, and H_2SO_4 dissociate completely in water, while weak acids dissociate only partially. Practically speaking, a weak acid is any acid that doesn't dissociate completely in water.

To measure the amount of dissociation occurring when a weak acid is in aqueous solution, chemists use a constant called the *acid dissociation constant* (K_a). K_a is a special variety of the equilibrium constant introduced in Chapter 14. As we explain in Chapter 14, the equilibrium constant of a chemical reaction is the concentration of products over the concentration of reactants and indicates the balance between products and reactants in a reaction. The acid dissociation constant is simply the equilibrium constant of a reaction in which an acid is mixed with water and from which the water concentration has been removed. This is done because the concentration of water is a constant in dilute solutions, and a better indicator of acidity is the concentration of the dissociated products divided by the concentration of the acid reactant. The general form of the acid dissociation constant is therefore

$$K_a = \frac{[H_3O^+]\times[A^-]}{[HA]}$$

Where [HA] is the concentration of the acid before it loses its hydrogen, and [A] is the concentration of its conjugate base. Notice that the concentration of the hydronium ion (H_3O^+) is used in place of the concentration of H^+, which we use to describe acids earlier in this chapter. In truth, they are one and the same. Generally speaking, H^+ ions in aqueous solution will be caught up by atoms of water in solution, making H_3O^+ ions.

A similar situation exists for bases. Strong bases such as KOH, NaOH, and $Ca(OH)_2$ dissociate completely in water. Weak bases don't dissociate completely in water, and their strength is measured by the *base dissociation constant,* or K_b.

$$K_b = \frac{[OH^-]\times[B^+]}{[BOH]}$$

Here, BOH is the base, and B is its conjugate acid. This can also be written in terms of the acid and base:

$$K_b = \frac{[OH^-]\times[HA]}{[A^-]}$$

In problems where you're asked to calculate K_a or K_b, you'll generally be given the concentration, or molarity, of the original acid or base and the concentration of its conjugate *or* hydronium/hydroxide, but rarely both. Dissociation of a single molecule of acid involves the splitting of that acid into one molecule of its conjugate base and one hydrogen ion, while dissociation of a base always involves the splitting of that base into one molecule of the conjugate acid and one hydroxide ion. For this reason, the concentration of the conjugate and the concentration of the hydronium or hydroxide ion in any dissociation are equal, so you only need one to know the other. You may also be given the pH and be asked to figure out the $[H^+]$ (equivalent to $[H_3O^+]$) or $[OH^-]$. K_a and K_b are also constants at constant temperature.

 Q. Write a general expression for the acid dissociation constant of the following reaction, a dissociation of ethanoic acid:

$$CH_3COOH + H_2O \rightarrow H_3O^+ + CH_3COO^-$$

Then calculate its actual value if $[CH_3COOH] = 2.34 \times 10^{-4}$ and $[CH_3COO^-] = 6.51 \times 10^{-5}$.

A. Your general expression should be

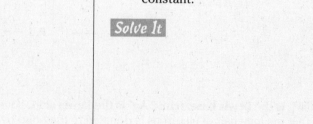

$K_a = \dfrac{[H_3O^+] \times [CH_3COO^-]}{[CH_3COOH]}$; the actual value is 1.80×10^{-5}. You're given $[CH_3COO^-] = 6.51 \times 10^{-5}$, and $[H_3O^+]$ must be the same. Plugging all known values into the expression for K_a yields

$$K_a = \frac{[6.51 \times 10^{-5}] \times [6.51 \times 10^{-5}]}{[2.34 \times 10^{-4}]} = 1.80 \times 10^{-5}$$

This is a very small K_a and thus a very weak acid.

---

8. Calculate the K_a for a 0.50M solution of benzoic acid, C_6H_5COOH, if the $[H^+]$ concentration is 5.6×10^{-3}.

Solve It

9. The pH of a 0.75M solution of HCOOH is 1.93. Calculate the acid dissociation constant.

Solve It

---

10. If you have a 0.2M solution of ammonia, NH_3, that has a K_b of 1.80×10^{-5}, what is the pH of that solution?

Solve It

Answers to Questions on Acids and Bases

The following are the answers to the practice problems presented in this chapter.

1   In this reaction, HCl donates a proton to H_2O, making it the Brønsted-Lowry acid. Water, which accepts the proton, is the Brønsted-Lowry base. This makes H_3O^+ the conjugate acid and Cl^- the conjugate base. Water can act as the base in this reaction and as an acid in the example problem because it's composed of both a hydrogen ion and a hydroxide ion and can therefore either accept or donate a proton.

2   For Arrhenius acids, remember to track the movement of H^+ and OH^- ions. If the reaction yields an OH^- product, the substance is a base, while an H^+ product reveals the substance to be an acid.

     a. Arrhenius base. NaOH dissociates in water to form OH^- ions, making it an Arrhenius base.

     b. Arrhenius acid. HF dissociates in aqueous solution to form H^+ ions, making it an Arrhenius acid.

3   To identify Lewis acids and bases, track the movement of electron pairs. Draw a Lewis dot structure to locate the atom with a lone pair available to donate. This is the Lewis base.

     a. H_2O is the Lewis base, while Cr^{3+} is the Lewis acid. Your Lewis dot structure should show that the lone pair of electrons is donated to the bond by H_2O.

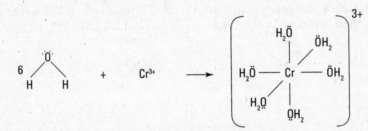

     b. NH_3 is the Lewis base, while Ag^+ is the Lewis acid. Your Lewis dot structure should show that the lone pair of electrons is donated to the bond by NH_3.

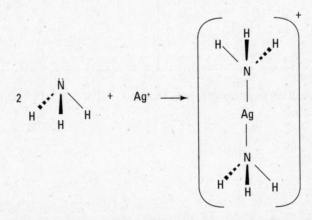

     c. The negative charge on the chlorine indicates that it's electron rich, making it the Lewis base. $HgCl_2$ is, therefore, the Lewis acid.

4 To calculate the pH values in this problem, use the equation pH = $-\log$ [H$^+$] and plug in all known values. If you're given [OH$^-$] instead of [H$^+$], simply calculate the pOH (using the equation pOH = $-\log$[OH$^-$]) instead and then subtract that number from 14 to give you the pH.

a. pH = 13. Because the coefficient on the H$^+$ concentration is 1, the pH is simply the exponent.

b. pH = 8.8. Plug the given [H$^+$] into the pH equation, giving pH = $-\log (1.58 \times 10^{-9})$ = 8.8

c. pH = 7.3. First, use the OH$^-$ concentration to calculate the pOH. pOH = $-\log(2 \times 10^{-7})$ = 6.7. Then use the relationship between pH and pOH (14 $-$ pOH = pH) to determine the pH: 14 $-$ 6.7 = 7.3.

d. pH = 7. Because the coefficient of this OH$^-$ concentration is 1, you can read the pOH directly from the exponent to get pOH = 7. Then use the relationship between pH and pOH (14 $-$ pOH = pH) to determine the pH: 14 $-$ 7 = 7.

5 To calculate pOH, simply take the negative logarithm of the OH$^-$ concentration. If you're given the H$^+$ concentration, use it to calculate the pH and then subtract that value from 14 to yield the pOH.

a. pOH = 6.3. First, use the H$^+$ concentration to calculate the pH. pH = $-\log(2 \times 10^{-8})$ = 7.7. Then use the relationship between pH and pOH (14 $-$ pH = pOH) to determine the pOH: 14 $-$ 7.7 = 6.3.

b. pOH = 10.3. Here, you can simply use the OH$^-$ concentration to calculate the pOH directly. pOH = $-\log (5.1 \times 10^{-11})$ = 10.3

6 To determine acidity, you need to calculate pH.

a. Basic. Use [OH$^-$] to determine pOH. pOH = $-\log(2.5 \times 10^{-7})$ = 6.6. Subtract this value from 14 to get a pH of 7.4. This solution is very slightly greater than 7 and, therefore, slightly basic.

b. Acidic. Use [OH$^-$] to determine pOH. pOH = $-\log(3.1 \times 10^{-13})$ = 12.5. Subtract this value from 14 to get a pH of 1.5. This solution is significantly smaller than 7 and, therefore, quite acidic.

c. Acidic. Use the [H$^+$] concentration to determine pH. pH = $-\log(4.21 \times 10^{-5})$ = 4.4.

d. Basic. Use the [H$^+$] concentration to determine pH. pH = $-\log(8.9 \times 10^{-10})$ = 9.1.

7 In this problem, use the formula [H$^+$] = 10^{-pH}. If you're given the pOH instead of the pH, begin by subtracting the pOH from 14 to give you the pH and then plug that number into the formula.

a. 5.0×10^{-4}. Use the pH to calculate the H$^+$ equation using the equation [H$^+$] = 10^{-pH}. [H$^+$] = $10^{-3.3}$ = 5.0×10^{-4}

b. 2.04×10^{-8}. Use the pH to calculate the H$^+$ equation using the equation [H$^+$] = 10^{-pH}. [H$^+$] = $10^{-7.69}$ = 2.04×10^{-8}

c. 1.62×10^{-4}. Begin by using the pOH to calculate the pH using the equation 14 $-$ pOH = pH. 14 $-$ 10.21 = 3.79 = pH. Then use the pH to calculate the H$^+$ equation using the equation [H$^+$] = 10^{-pH}. [H$^+$] = $10^{-3.79}$ = 1.62×10^{-4}

d. 1.82×10^{-13}. Begin by using the pOH to calculate the pH using the equation 14 $-$ pOH = pH. 14 $-$ 1.26 = 12.74 = pH. Then use the pH to calculate the H$^+$ equation using the equation [H$^+$] = 10^{-pH}. [H$^+$] = $10^{-12.74}$ = 1.82×10^{-13}

8 **6.3×10^{-5}.** The molarity in this problem tells you that the concentration of benzoic acid is 5.0×10^{-1}. You also know that the concentration of benzoic acid's conjugate base is the same as the given H^+ concentration. All that remains is to write an equation for the acid dissociation constant and plug in these concentrations.

$$K_a = \frac{[H_3O^+] \times [C_6H_5COO^-]}{[C_6H_5COOH]} = \frac{[5.6 \times 10^{-3}] \times [5.6 \times 10^{-3}]}{[5.0 \times 10^{-1}]} = 6.3 \times 10^{-5}$$

9 **1.6×10^{-4}.** You need to use the pH to calculate the $[H^+]$. $[H^+] = 10^{-pH}$. $[H^+] = 10^{-1.93} = 1.1 \times 10^{-2}$, which must also be equal to the concentration of the conjugate base. All that remains is to write an equation for the acid dissociation constant and plug in these concentrations.

$$K_a = \frac{[H_3O^+] \times [HCOO^-]}{[HCOOH]} = \frac{[1.1 \times 10^{-2}] \times [1.1 \times 10^{-2}]}{[7.5 \times 10^{-1}]} = 1.6 \times 10^{-4}$$

10 **pH = 8.6.** Begin by considering the K_b equation:

$$K_b = \frac{[OH^-] \times [NH_4^+]}{[NH_3]}.$$

Because $[OH^-] = [NH_4^+]$, you can rewrite this as $K_b = \frac{[OH^-]^2}{[NH_3]}$.

Solving this equation for $[OH^-]$ yields $[OH^-] = \sqrt{K_b \times [NH_3]}$.

Plugging in the known values of K_b and $[NH_3]$ into this equation yields

$$[OH^-] = \sqrt{1.8 \times 10^{-5} \times 0.2} = 3.6 \times 10^{-6}$$

Use this to solve for the pOH.

pOH = $-\log (3.6 \times 10^{-6}) = 5.4$.

The final step is to solve for pH by subtracting this number from 14.

$14 - 5.4 = 8.6$

Chapter 17

Achieving Neutrality with Equivalents, Titration, and Buffers

● ●

In This Chapter

▶ Surveying the superior neutralizing ability of certain acids and bases

▶ Deducing the concentration of a mystery acid or base through titration

▶ Working with buffered solutions

▶ Measuring the solubility of saturated solutions of salts with K_{sp}

● ●

Chapter 16 gives you the ins and outs of acids and bases, but treats each separately. In the real world of chemistry, however, acids and bases often meet in solution, and when they do, they're drawn to one another. These unions of acid and base are called *neutralization reactions* because the low pH of the acid and the high pH of the base essentially cancel one another out, resulting in a neutral solution. In fact, when a hydroxide-containing base reacts with an acid, the products are simply an innocuous salt and water.

Although strong acids and bases have their uses, the prolonged presence of a strong acid or base in an environment not equipped to handle it can be very damaging. In the laboratory, for example, you need to handle strong acids and bases carefully, and deliberately perform neutralization reactions when appropriate. A lazy chemistry student will get a brow beating from her hawk-eyed teacher if she attempts to dump a concentrated acid or base down the laboratory sink. Doing so can damage the pipes and is generally unsafe. Instead, a responsible chemist will neutralize acidic laboratory waste with a base such as baking soda, and will neutralize basic waste with an acid. Doing so makes a solution perfectly safe to dump down the drain and often results in the creation of a satisfyingly sizzly solution while the reaction is occurring.

In this chapter, we explore issues that arise when acids meet up with bases. First, not all acids and bases are created equally — some compounds carry more acid or base *equivalents* than others do. To account for these differences, we use *normality,* a measure of the effective amounts of acid or base a compound produces in solution. Connected to the ideas of equivalents and normality is *titration,* the process by which chemists add acids to bases (or vice versa) a little bit at a time, gradually using up acid and base equivalents as the two neutralize each other. We show you how to use titration to figure out the concentration of an unknown acid. Then, we describe *buffer solutions,* mixtures that contain both acid and base forms of the same compounds and serve to maintain the pH of the solution even when extra acid or base is added. Finally, because salts are produced when acids react with bases, we discuss the solubility product, K_{sp}, a number that tells you how soluble a salt is in solution.

At heart, neutralization reactions in which the base contains a hydroxide ion are simple double replacement reactions of the form $HA + BOH \rightarrow BA + H_2O$ (or in other words, an acid reacts with a base to form a salt and water). You're asked to write a number of such reactions in this chapter, so be sure to review double replacement reactions and balancing equations in Chapter 8 before you delve into the new and exciting world of neutralization.

Examining Equivalents and Normality

Unlike people, not all acids and bases are created equally. Some have an innate ability to neutralize more effectively than others. Consider hydrochloric acid (HCl) and sulfuric acid (H_2SO_4), for example. If you mixed $1M$ sodium hydroxide (NaOH) together with $1M$ hydrochloric acid, you'd need to add equal amounts of each to create a neutral solution. If you mixed sodium hydroxide with sulfuric acid, however, you'd need to add twice as much sodium hydroxide as sulfuric acid. Why this blatant inequality of acids? The answer lies in the balanced neutralization reactions for both acid/base pairs.

$$HCl + NaOH \rightarrow NaCl + H_2O$$

$$H_2SO_4 + 2NaOH \rightarrow Na_2SO_4 + 2H_2O$$

The coefficients in the balanced equations are the key to understanding this inequality. To balance the equation, the coefficient 2 needs to be added to sodium hydroxide, indicating that 2 moles of it must be present to neutralize 1 mole of sulfuric acid. On a molecular level, this happens because sulfuric acid has two acidic hydrogen atoms to give up, and the single hydroxide in a molecule of sodium hydroxide can only neutralize one of those two acidic hydrogens to form water. Therefore, 2 moles of sodium hydroxide are needed for every 1 mole of sulfuric acid. Hydrochloric acid, on the other hand, has only one acidic hydrogen to contribute, so it can be neutralized by an equal amount of sodium hydroxide, which has only one hydroxide to contribute to neutralization.

The number of moles of an acid or base multiplied by the number of hydrogens or hydroxides that a molecule has to contribute in a neutralization reaction is called the number of *equivalents* of that substance. Basically, the number of *effective* neutralizing moles available determines the ratio of acid to base in a neutralization reaction.

You may also be asked to calculate the *normality (N)* of an acid or base, which is simply the number of equivalents divided by the volume in liters.

$$\text{normality} = \frac{\text{equivalents}}{\text{volume in L}}$$

The normality of an acidic or basic solution is often given in place of the molarity of that solution. In many ways, the two are quite similar. Normality, however, unlike molarity, takes equivalents into account. Mixing equal amounts of acidic and basic solutions of equal normality always results in a neutral solution, while the same can't be said of solutions of equal molarity. (Flip to Chapter 12 for an introduction to molarity.)

You may also find the equation $N_A \times V_A = N_B \times V_B$ equating the volumes and normalities of a neutral acid/base mixture useful. Normality multiplied by volume is, after all, just the number of equivalents, and equal numbers of equivalents of acid and base must by definition neutralize one another.

Q. Your careless lab partner has just dumped 24 mL of 0.8*M* magnesium hydroxide, Mg(OH)$_2$, into your lab sink. Luckily, a plug in the bottom of the sink is keeping it from entering the drain. You must neutralize the base before you can pull the plug. The only acid you have at your disposal is 0.4*M* phosphoric acid (H$_3$PO$_4$). How many milliliters should you add to the sink before making your lab partner reach her hand in to pull the plug (assuming, of course, that you don't wish her any harm)?

A. **32 mL H$_3$PO$_4$.** The first step to solving this problem is to determine the number of equivalents of Mg(OH)$_2$ in the lab sink. To do that, you first need the number of moles of base. Because you're given molarity and volume, you simply multiply the volume in liters by the molarity.

$$\frac{0.024L}{1} \times \frac{0.8\ mol}{1L} = 0.0192\ mol\ Mg\big(OH\big)_2$$

To find the number of equivalents in the sink, you have to multiply this number by 2 to compensate for the two hydroxide ions available for neutralization per each mole of magnesium hydroxide, giving you 0.0384 equivalents Mg(OH)$_2$.

To neutralize this, you need an equal number of equivalents of phosphoric acid. You know you need 0.0384 equivalents of H$_3$PO$_4$ and that there are 3 equivalents per mole of H$_3$PO$_4$ (because there are three hydrogen atoms), so you need 0.0128 mol of H$_3$PO$_4$ to neutralize the base. Dividing this value by the molarity of your solution (or, equivalently, multiplying by its reciprocal) gives you the proper number of liters to add.

$$\frac{0.0128\ mol}{1} \times \frac{1L}{0.4\ mol} = 0.032L,\ or\ 32\ mL\ H_3PO_4$$

---

1. Write a balanced neutralization reaction for a reaction between hydrofluoric acid, HF, and calcium hydroxide, Ca(OH)$_2$. How many equivalents of each are present?

Solve It

2. How many moles of potassium hydroxide, KOH, do you need to neutralize 3 moles of nitric acid, HNO$_3$?

Solve It

3. What is the normality of a solution containing 1.5M H$_2$SO$_4$?

4. How many milliliters of 1.5M phosphoric acid, H$_3$PO$_4$, do you need to neutralize 20 mL of 2N potassium hydroxide, KOH?

Concentrating on Titration to Figure Out Molarity

Imagine you're a newly hired laboratory assistant who's been asked to alphabetize the chemicals on the shelves of a chemistry laboratory during a lull in experimenting. As you reach for the bottle of sulfuric acid, your first-day jitters get the better of you, and you knock over the bottle. Some careless chemist failed to screw the cap on tightly! You quickly neutralize the acid with a splash of baking soda and wipe up the now nicely neutral solution. As you pick up the bottle, however, you notice that the spilled acid burned away most of the label! You know it's sulfuric acid, but there are several different concentrations of sulfuric acid on the shelves, and you don't know the molarity of the solution in this bottle. Knowing that your boss will surely blame you if she sees the damaged bottle, and not wanting to get sacked on your very first day, you quickly come up with a way to determine the molarity of the solution and save your job.

You know that the bottle contains sulfuric acid of a mystery concentration, and you notice bottles of 1M sodium hydroxide, a strong base, and phenolphthalein, a pH indicator, among the chemicals on the shelves. You measure a small amount of the mystery acid into a beaker and add a little phenolphthalein. You reason that if you drop small amounts of sodium hydroxide into the solution until the phenolphthalein indicates that the solution is neutral by turning the appropriate color, you'll be able to figure out the acid's concentration.

You can do this by making a simple calculation of the number of moles of sodium hydroxide you have added, and then reasoning that the mystery acid must have an equal number of equivalents to have been neutralized. This then leads to the number of moles of acid, and that, in turn, can be divided by the volume of acid you added to the beaker to get the molarity. Whew! You relabel the bottle and rejoice in the fact that you can come in and do menial labor in the lab again tomorrow.

This process is called a *titration,* and it's often used by chemists to determine the molarity of acids and bases. In a titration calculation, you generally know the identity of an acid or base of unknown concentration, and the identity and molarity of the acid or base which you're going to use to neutralize it. Given this information, you then follow six simple steps:

1. **Measure out a small volume of the mystery acid or base.**

2. **Add a pH indicator such as phenolphthalein.**

Be sure to take note of the color that will indicate that the solution has reached neutrality.

3. **Neutralize.**

Slowly add the acid or base of known concentration into the solution until the indicator shows that it's neutral, keeping careful track of the volume added.

4. **Calculate the number of moles added.**

Multiply the number of liters of acid or base added by the molarity of that acid or base to get the number of moles added.

5. **Account for equivalents.**

Determine how many moles of the mystery substance being neutralized are present using equivalents. How many moles of your acid do you need to neutralize one mole of base or vice versa?

6. **Solve for molarity.**

Divide the number of moles of the mystery acid or base by the number of liters measured out in Step 1, giving you the molarity.

The titration process is often visualized using a graph showing concentration of base on one axis and concentration of acid on the other as in Figure 17-1. The interaction of the two traces out a *titration curve,* which has a characteristic *s* shape.

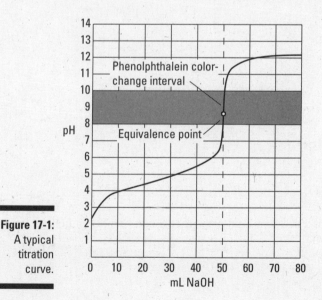

Figure 17-1:
A typical titration curve.

Q. If the laboratory assistant had to add 10 mL of $1M$ sodium hydroxide to neutralize 5 mL of the sulfuric acid in his impromptu titration, what did he end up writing on the label for the concentration of the sulfuric acid in the bottle?

A. **$1M$ H_2SO_4.** The problem tells us that the volume from Step 1 of the titration process is 5 mL and that the volume of base from Step 3 is 10 mL. In Step 4, you must calculate the number of moles of sodium hydroxide added by multiplying the volume in liters (0.01L) by the molarity

($1M$) to give 0.01 mol NaOH. Because one NaOH has only one hydroxide ion to contribute to neutralization, there are also 0.01 equivalents of NaOH present. You need an equal number of equivalents of H_2SO_4 to complete the neutralization. But H_2SO_4 has two equivalents to contribute per mole, so you only need half as many moles of the acid, or 0.005 mol. The final step is to divide this value by the volume of acid in liters added to get a molarity of 1. The assistant should have labeled the bottle as $1M$ H_2SO_4.

---

5. If, in doing a titration of a solution of calcium hydroxide of unknown concentration, a student adds 12 mL of $2M$ HCl (hydrochloric acid) and finds that the molarity of the base is $1.25M$, how much $Ca(OH)_2$ must the student have measured out at the start of the titration?

Solve It

6. Titration shows that a 5 mL sample of nitrous acid, HNO_2, has a molarity of 0.5. If 8 mL of magnesium hydroxide, $Mg(OH)_2$, were added to the acid to accomplish the neutralization, what must the molarity of the base have been?

Solve It

---

7. How much could a chemist find out about a mystery acid or base through titration if neither its identity nor its concentration were known?

Maintaining Your pH with Buffers

You may have noticed that the titration curve shown in Figure 17-1 has a flattened area in the middle where pH doesn't change significantly, even when you add a conspicuous amount of base. This region is called a *buffer region*.

Certain solutions, called *buffered solutions,* resist changes in pH like a stubborn child resists eating her Brussels sprouts: steadfastly at first, but choking them down reluctantly if enough pressure is applied (such as the threat of no dessert). Although buffered solutions maintain their pH very well when small amounts of acid or base are added to them or the solution is diluted, they can only withstand the addition of a certain amount of acid or base before becoming overwhelmed.

Buffers are most often made up of a weak acid and its conjugate base, though they can also be made of a weak base and its conjugate acid. (Conjugate bases and acids are the products in acid-base reactions; see Chapter 16 for details.) A weak acid in aqueous solution will be partially dissociated, and the amount of dissociation depends on its pK_a value (the negative logarithm of its acid dissociation constant). The dissociation will be of the form $HA + H_2O \rightarrow H_3O^+ + A^-$, where A^- is the conjugate base of the acid HA. The acidic proton is taken up by a water molecule, forming hydronium. If HA were a strong acid, 100 percent of the acid would become H_3O^+ and A^-, but because it's a weak acid, only a fraction of the HA dissociates, and the rest remains HA.

The Ka (acid dissociation constant; see Chapter 16) of this reaction is defined by

$$K_a = \frac{\left[H_3O^+\right]\left[A^-\right]}{\left[HA\right]}$$

Solving this equation for the $[H_3O^+]$ concentration allows you to devise a relationship between the $[H_3O^+]$ and the K_a of a buffer.

$$\left[H_3O^+\right] = K_a \times \frac{\left[HA\right]}{\left[A^-\right]}$$

Taking the negative logarithm of both sides of the equation and manipulating logarithm rules yields an equation called the *Henderson-Hasselbach equation* that relates the pH and the pK_a.

$$pH = pK_a + \log\left(\frac{\left[A^-\right]}{\left[HA\right]}\right)$$

This equation can be manipulated using logarithm rules to get $\frac{\left[A^-\right]}{\left[HA\right]} = 10^{pH-pK_a}$, which may be more useful in certain situations.

The very best buffers and those best able to withstand the addition of both acid and base are those for which [HA] and [A⁻] are approximately equal. When this occurs, the logarithmic term in the Henderson-Hasselbach equation disappears, and the equation becomes $pH = pK_a$. When creating a buffered solution, chemists therefore choose an acid that has a pK_a close to the desired pH.

If you add a strong base such as sodium hydroxide (NaOH) to this mixture of dissociated and undissociated acid, its hydroxide is absorbed by the acidic proton, replacing the exceptionally strong base OH⁻ with a relatively weak base A⁻, and minimizing the change in pH.

$$HA + OH^- \rightarrow H_2O + A^-$$

This causes a slight excess of base in the reaction, but doesn't affect pH significantly. You can think of the undissociated acid as a reservoir of protons that are available to neutralize any strong base that may be introduced to the solution. As we explain in Chapter 14, when a product is added to a reaction, the equilibrium in the reaction changes to favor the reactants or to "undo" the change in conditions. Because this reaction generates A⁻, the acid dissociation reaction happens less frequently as a result, further stabilizing the pH.

When a strong acid, such as hydrochloric acid (HCl), is added to the mixture, its acidic proton is taken up by the base A⁻, forming HA.

$$H^+ + A^- \rightarrow HA$$

This causes a slight excess of acid in the reaction but doesn't affect pH significantly. It also shifts the balance in the acid dissociation reaction in favor of the products, causing it to happen more frequently and re-creating the base A⁻.

The addition of acid and base and their effect on the ratio of products and reactants is summarized in Figure 17-2.

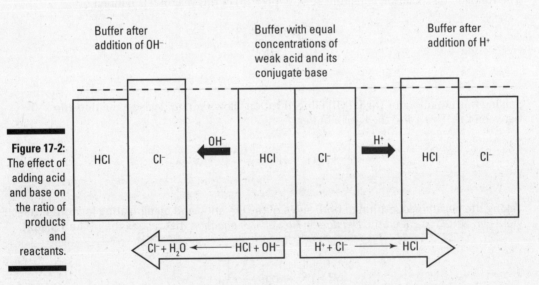

Figure 17-2: The effect of adding acid and base on the ratio of products and reactants.

Buffers have their limits, however. The acid's proton reservoir, for example, can only compensate for the addition of a certain amount of base before it runs out of protons that can neutralize free hydroxide. At this point, a buffer has done all it can do, and the titration curve resumes its steep upward slope.

Q. Consider a buffered solution that contains the weak acid ethanoic acid ($K_a = 1.8 \times 10^{-5}$) and its conjugate base, ethanoate. If the concentration of the solution is $0.5M$ with respect to ethanoic acid and $0.3M$ with respect to ethanoate, what is the pH of the solution?

A. **pH = 4.5.** This is a simple application of the Henderson-Hasselbach equation; remember to take the negative logarithm of K_a to get pK_a. Plugging in known values yields

$$pH = pK_a + \log\left(\frac{[A^-]}{[HA]}\right) = -\log(1.8 \times 10^{-5}) + \log\left(\frac{[0.3]}{[0.5]}\right) = 4.5$$

---

8. Ethanoic acid would make an ideal buffer to maintain what approximate pH?

9. Describe the preparation of 1,000 mL of a $0.2M$ carbonic acid buffer of pH = 7.0. Assume the pK_a of carbonic acid is 6.8.

Solve It

Measuring Salt Solubility: K_{sp}

In chemistry, a "salt" is not necessarily the substance you sprinkle on French fries, but has a much broader definition. Rather, a *salt* is any substance that is a combination of an anion (an atom with a negative charge) and a cation (an atom with a positive charge) and is created in a neutralization reaction. Salts, therefore, tend to dissociate in water. The degree of dissociation possible — in other words, the solubility of the salt — varies greatly from one salt to another.

Chemists use a quantity called the *solubility product constant* or K_{sp} to compare the solubilities of salts. K_{sp} is calculated in much the same way as an equilibrium constant (K_{eq}). The product concentrations are multiplied together, each raised to the power of its coefficient in the balanced dissociation equation. There is one key difference, however, between a K_{sp} and a K_{eq}. K_{sp} is a quantity specific to a *saturated* solution of salt, so the concentration of the undissociated salt reactant has absolutely no bearing on its value. If the solution is saturated, then the amount of possible dissociation is at its maximum, and any additional solute added merely settles on the bottom.

Q. Write a formula for the solubility product constant of the reaction $CaF_2 \rightarrow Ca^{2+} + 2F^-$.

A. $K_{sp} = [Ca^{2+}] \times [F^-]^2$. The solubility product constant is constructed by raising the concentrations of the two products to the power of their coefficients, so $K_{sp} = [Ca^{2+}] \times [F^-]^2$.

10. Write the solubility product dissociation constants for silver (I) chromate, Ag_2CrO_4, and strontium sulfate, $SrSO_2$.

Solve It

11. If the K_{sp} of silver chromate is 1.1×10^{-12} and the silver ion concentration in the solution is $0.0005M$, what is the chromate concentration?

Solve It

Answers to Questions on Neutralizing Equivalents

The following are the answers to the practice problems presented in this chapter.

1 **2HF + Ca(OH)$_2$ → CaF$_2$ + 2H$_2$O.** The balanced neutralization reaction tells you that there are 2 equivalents of CaOH$_2$ per mole and 1 equivalent of HF per mole.

2 **3 mol.** Nitric acid (HNO$_3$) is a monoprotic acid (an acid with a single hydrogen atom) with 1 equivalent per mole, while potassium hydroxide (KOH) also has 1 equivalent per mole. So you need equal amounts of each to achieve neutralization.

3 **3N H$_2$SO$_4$.** Normality is equivalents per liter. Because volume is already built into molarity, you simply need to multiply the molarity by the number of equivalents in 1 mol of H$_2$SO$_4$ (2 equivalents, one for each hydrogen).

$$\frac{1.5 \text{ mol H}_2\text{SO}_4}{1} \times \frac{2 \text{ equivalents H}_2\text{SO}_4}{1 \text{ mol H}_2\text{SO}_4} = 3\text{N H}_2\text{SO}_4$$

4 **8.9 mL.** The quickest way to arrive at the answer is to solve for the normality of the phosphoric acid so you can use the equation $N_A \times V_A = N_B \times V_B$. Calculate this normality with the equation

$$\frac{1.5 \text{ mol H}_3\text{PO}_4}{1} \times \frac{3 \text{ equivalents H}_3\text{PO}_4}{1 \text{ mol H}_3\text{PO}_4} = 4.5\text{N}$$

Solve the equation $N_A \times V_A = N_B \times V_B$ for V_A and plug in all known values to get your answer.

$$V_A = \frac{N_B \times V_B}{N_A} = \frac{2\text{N} \times 20 \text{ mL}}{4.5\text{N}} = 8.9 \text{ mL}$$

5 **9.6 mL Ca(OH)$_2$.** Begin by finding the number of moles of HCl by multiplying the molarity ($2M$) by the volume (0.012L) to get 0.024 mol HCl. Next, calculate the number of moles of Ca(OH)$_2$ needed to neutralize this amount using equivalents:

$$\frac{0.024 \text{ mol HCl}}{1} \times \frac{1 \text{ mol Ca(OH)}_2}{2 \text{ mol HCl}} = 0.012 \text{ mol Ca(OH)}_2$$

Divide this number by the molarity to get the volume of base added, giving 0.0096L Ca(OH)$_2$.

6 **0.16M Mg(OH)$_2$.** This problem gives you Step 6 in the titration procedure and asks you to back-solve for the molarity of the base. Start by finding the number of moles of acid present in the solution by multiplying the molarity ($0.5M$) by the volume (0.005L), giving you 0.0025 mol. Next, determine the number of equivalents of magnesium hydroxide needed to neutralize 0.0025 mol HNO$_2$ by examining the balanced neutralization reaction.

2HNO$_2$ + Mg(OH)$_2$ → Mg(NO$_2$)$_2$ + 2H$_2$O

$$\frac{0.0025 \text{ mol HNO}_2}{1} \times \frac{1 \text{ mol Mg(OH)}_2}{2 \text{ mol HNO}_2} = 1.25 \times 10^{-3} \text{ mol Mg(OH)}_2$$

Divide this by the volume of base (8 mL, or 0.008L) added to get a molarity of 0.16M Mg(OH)$_2$.

7 Without an identity or a concentration, a chemist could still determine the number of equivalents per liter of the mystery acid or base. If he titrates an extra-mysterious acid with a base of known concentration until he achieves neutrality, then he knows the number of equivalents of acid in the solution (equal to the number of equivalents of base added). This information may even allow him to guess at its identity.

8 **pH of 4.7.** Ethanoic acid, also known as acetic acid or vinegar, would be an ideal buffer for a pH close to its pK_a value. You're given the K_a of ethanoic acid in the example problem (it's $K_a = 1.8 \times 10^{-5}$), so simply take the negative logarithm of that value to get your pK_a value and therefore your pH.

9 You're given a pH and a pK_a, which suggests that you need to use the Henderson-Hasselbach equation. You're given the total concentration of acid and conjugate base, but you don't know either of the concentrations in the equation individually, so begin by solving for their ratio.

$$\frac{\left[A^-\right]}{\left[HA\right]} = 10^{pH-pK_a} = 10^{7-6.8} = 1.6$$

This tells you that there are 1.6 mol of base for every 1 mol of acid. Expressing the amount of base as a fraction of the total amount of acid and base therefore gives you

$$\frac{\left[A^-\right]}{\left[HA\right]+\left[A^-\right]} = \frac{1.6}{1+1.6} = 0.62$$

Multiply this number by the molarity of the solution ($0.2M$) to give you the molarity of the basic solution, or $0.12M$. The molarity of the acid is therefore $0.2M - 0.12M = 0.08M$.

This means that to prepare the buffer you must take 0.2 mol of carbonic acid and add it to somewhat less than 1,000 mL of water (say 800 mL). To this you must add enough of a strong base such as sodium hydroxide (NaOH) to force the proper proportion of the carbonic acid solution to dissociate into its conjugate base. Because you want to achieve a base concentration of $0.12M$, you should add 120 mL of NaOH. Finally, you should add enough water to achieve your final volume of 1,000 mL.

10 For silver (I) chromate: $K_{sp} = [Ag^+]^2 \times [CrO_4^{2-}]$; for strontium sulfate: $K_{sp} = [Sr^{2+}] \times [SO_2^{2-}]$.

To determine the solubility product constants of these solutions, you first need to write an equation for their dissociation in water (see Chapter 8 for details).

$Ag_2CrO_4 \rightarrow 2Ag^+(aq) + CrO_4^{2-}(aq)$

$SrSO_2 \rightarrow Sr^{2+}(aq) + SO_2^{2-}(aq)$

Raising the concentration of the products to the power of their coefficients in these balanced reactions yields the K_{sp} for each.

$K_{sp} = [Ag^+]^2 \times [CrO_4^{2-}]$

$K_{sp} = [Sr^{2+}] \times [SO_2^{2-}]$

11 4×10^{-6}. You wrote an expression for the K_{sp} of silver (I) chromate in Problem 10. Solve the equation for the chromate concentration by dividing K_{sp} by the silver ion concentration.

$$\left[CrO_4^{2-}\right] = \frac{K_{sp}}{\left[Ag^+\right]^2} = \frac{1.1 \times 10^{-12}}{\left(5 \times 10^{-4}\right)^2} = 4 \times 10^{-6}$$

Chapter 18

Accounting for Electrons in Redox

. . . .

In This Chapter

▶ Keeping an eye on electrons by using oxidation numbers

▶ Balancing redox reactions in the presence of acid

▶ Balancing redox reactions in the presence of base

. . . .

*I*n chemistry, electrons get all the action. Among other things, electrons can transfer between reactants during a reaction. Reactions like these are called *oxidation-reduction* reactions, or *redox* reactions for brevity. Redox reactions are as important as they are common. But they're not always obvious. In this chapter, you find out how to recognize redox reactions and how to balance the equations that describe them.

Keeping Tabs on Electrons with Oxidation Numbers

If electrons move between reactants during redox reactions, then it should be easy to recognize those reactions just by noticing changes in charge, right?

Sometimes.

The following two reactions are both redox reactions:

$$Mg(s) + 2H^+(aq) \rightarrow Mg^{2+}(aq) + H_2(g)$$

$$2H_2(g) + O_2(g) \rightarrow 2H_2O(g)$$

In the first reaction, it's obvious that electrons are transferred from magnesium to hydrogen. (There are only two reactants, and magnesium is neutral as a reactant but positively charged as a product.) In the second reaction, it's not so obvious that electrons are transferred from hydrogen to oxygen. We need a way to keep tabs on electrons as they transfer between reactants. In short, we need *oxidation numbers*.

Oxidation numbers are tools to keep track of electrons. Sometimes an oxidation number describes the actual charge on an atom. Other times an oxidation number describes an imaginary sort of charge — the charge an atom would have if all of its bonding partners left town, taking their own electrons with them. In ionic compounds, oxidation numbers come closest to describing actual atomic charge. The description is less apt in covalent compounds, in which electrons are less clearly "owned" by one atom or another. (See Chapter 6 for the basics on different types of compounds.) The point is this: Oxidation numbers are useful tools, but they aren't direct descriptions of physical reality.

Here are some basic rules for figuring out an atom's oxidation number:

- ✔ Atoms in elemental form have an oxidation number of 0. So, the oxidation number of both $Mg(s)$ and $O_2(g)$ is 0.

- ✔ Single-atom *(monatomic)* ions have an oxidation number equal to their charge. So, the oxidation number of $Mg^{2+}(aq)$ is +2 and the oxidation number of $Cl^-(aq)$ is –1.

- ✔ In a neutral compound, oxidation numbers add up to 0. In a charged compound, oxidation numbers add up to the compound's charge.

- ✔ In compounds, oxygen usually has an oxidation number of –2. An annoying exception is the peroxides, like H_2O_2, in which oxygen has an oxidation number of –1.

- ✔ In compounds, hydrogen has an oxidation number of +1 when it bonds to nonmetals (as in H_2O), and an oxidation number of –1 when it bonds to metals (as in NaH).

- ✔ In compounds:

    - Group IA atoms (alkali metals) have an oxidation number of +1.

    - Group IIA atoms (alkaline earth metals) have an oxidation number of +2.

    - In Group IIIA, Al and Ga atoms have an oxidation number of +3.

    - Group VIIA atoms (halogens) usually have an oxidation number of –1.

You can deduce the oxidation numbers of other atoms in compounds (especially those of transition metals) by taking into account these basic rules, as well as the compound's overall charge.

By applying these rules to chemical reactions, you can discern who supplies electrons to whom.

- ✔ The chemical species that loses electrons is *oxidized,* and acts as the *reducing agent* (or *reductant*).

- ✔ The species that gains electrons is *reduced,* and acts as the *oxidizing agent* (or *oxidant*).

All redox reactions have both an oxidizing agent and a reducing agent. Although a given beaker can contain many different oxidizing agent–reducing agent pairs, each pair constitutes its own redox reaction.

Oxidation may or may not involve bonding with oxygen, breaking bonds with hydrogen, or losing electrons — but oxidation always means an increase in oxidation number. Reduction may or may not involve bonding with hydrogen, breaking bonds with oxygen, or gaining electrons — but reduction always means a decrease in oxidation number.

EXAMPLE

Q. Use oxidation numbers to identify the oxidizing and reducing agents in the following chemical reaction:

$$Cr_2O_3(s) + 2Al(s) \rightarrow 2Cr(s) + Al_2O_3(s)$$

A. **$Cr_2O_3(s)$ is the oxidizing agent, and Al(s) is the reducing agent.** The keys here are to recall that atoms in elemental form (like solid Al and solid Cr) have oxidation numbers of 0, and that oxygen typically has an oxidation number of –2 in compounds. The oxidation number breakdown (shown in the following figure) reveals that Al(s) is oxidized to $Al_2O_3(s)$, and $Cr_2O_3(s)$ is reduced to Cr(s). So, Al(s) is the reducing agent, and $Cr_2O_3(s)$ is the oxidizing agent. You can tell that Al(s) is oxidized because the oxidation number of Al is 0 in Al(s) but is +2 in $Al_2O_3(s)$. You can tell that $Cr_2O_3(s)$ is reduced because the oxidation number of Cr is +2 in $Cr_2O_3(s)$ but is 0 in Cr(s).

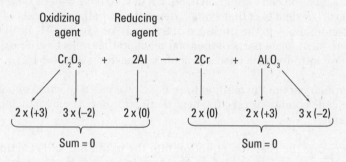

1. Use oxidation numbers to identify the oxidizing and reducing agents in the following chemical reaction:

$$MnO_2(s) + 2H^+(aq) + NO_2^-(aq) \rightarrow$$
$$NO_3^-(aq) + Mn^{2+}(aq) + H_2O(l)$$

Solve It

2. Use oxidation numbers to identify the oxidizing and reducing agents in the following chemical reaction:

$$H_2S(aq) + 2NO_3(aq) + 2H^+(aq) \rightarrow$$
$$S(s) + 2NO_2(g) + 2H_2O(l)$$

Solve It

Balancing Redox Reactions under Acidic Conditions

When you balance a chemical reaction equation, the primary concern is to obey the principle of Conservation of Mass — the total mass of the reactants must equal the total mass of the products. (See Chapter 8 if you need to review this process.) In redox reactions, you must obey a second principle as well: the *Conservation of Charge.* The total number of electrons lost must equal the total number of electrons gained. In other words, you can't just leave electrons lying around. The universe is finicky about that type of thing.

Sometimes simply balancing a redox reaction with an eye to mass results in a charge-balanced equation as well. Like a string of green lights when you're driving, that's a lovely thing when it occurs, but you can't count on it. So, it's best to have a go-to system for balancing redox reactions. The details of that system depend on whether the reaction occurs in acidic or basic conditions — in the presence of excess H^+ or excess OH^-. Both variations use *half-reactions,* incomplete parts of the total reaction that reflect either oxidation or reduction alone. (We introduce the basics of acids and bases in Chapter 16.)

Here's a summary of the method for balancing a redox reaction equation for a reaction under acidic conditions (see the next section for details on balancing a reaction under basic conditions):

1. **Separate the reaction equation into the oxidation half-reaction and the reduction half-reaction. Use oxidation numbers to identify these component half-reactions.**

2. **Balance the half-reactions separately, temporarily ignoring O and H atoms.**

3. **Turn your attention to the O and H atoms. Balance the half-reactions separately, using H_2O to add O atoms and using H^+ to add H atoms.**

4. **Balance the half-reactions separately for charge by adding electrons (e^-).**

5. **Balance the charge of the half-reactions with respect to each other by multiplying the reactions so that the total number of electrons is the same in each half-reaction.**

6. **Reunite the half-reactions into a complete redox reaction equation.**

7. **Simplify the equation by canceling items that appear on both sides of the arrow.**

Q. Balance the following redox reaction equation assuming acidic conditions:

$$NO_2^- + Br^- \rightarrow NO + Br_2$$

A. Simply go through the steps, 1 through 7.

Step 1: Divide the equation into half-reactions for oxidation and reduction by using oxidation numbers.

$$Br^- \rightarrow Br_2 \quad \text{(oxidation)}$$

$$NO_2^- \rightarrow NO \text{(reduction)}$$

Step 2: Balance the half-reactions, temporarily ignoring O and H.

$$2Br^- \rightarrow Br_2$$

$$NO_2^- \rightarrow NO$$

Step 3: Balance the half-reactions for O and H by adding H_2O and H^+, respectively.

$$2Br^- \rightarrow Br_2$$

$$2H^+ + NO_2^- \rightarrow NO + H_2O$$

Step 4: Balance the charge within each half-reaction by adding electrons (e⁻).

$2Br^- \rightarrow Br_2 + 2e^-$

$1e^- + 2H^+ + NO_2^- \rightarrow NO + H_2O$

Step 5: Balance the charge of the half-reactions with respect to each other.

$2Br^- \rightarrow Br_2 + 2e^-$

$2e^- + 4H^+ + 2NO_2^- \rightarrow 2NO + 2H_2O$

Step 6: Add the half-reactions, reuniting them within the total reaction equation.

$2e^- + 4H^+ + 2Br^- + 2NO_2^- \rightarrow Br_2 + 2NO + 2H_2O + 2e^-$

Step 7: Simplify by canceling items that appear on both sides of the equation.

$4H^+ + 2Br^- + 2NO_2^- \rightarrow Br_2 + 2NO + 2H_2O$

---

3. Balance the following redox reaction equation assuming acidic conditions:

$Fe^{3+} + SO_2 \rightarrow Fe^{2+} + SO_4^{2-}$

Solve It

4. Balance the following redox reaction equation assuming acidic conditions:

$Ag^+ + Be \rightarrow Ag + Be_2O_3^{2-}$

Solve It

---

5. Balance the following redox reaction equation assuming acidic conditions:

$Cr_2O_7^{2-} + Bi \rightarrow Cr^{3+} + BiO^+$

Solve It

Balancing Redox Reactions under Basic Conditions

It's true: Balancing redox equations can involve quite a lot of bookkeeping. Not much can be done to remedy that unfortunate fact. But here's the good news — the process for balancing redox equations under basic conditions is 90 percent identical to the one used for balancing under acidic conditions in the previous section. In other words, master one and you've mastered both.

Here's how easy it is to adapt your balancing method for basic conditions:

✔ Perform Steps 1–7 as described in the previous section for balancing under acidic conditions.

✔ Observe where H^+ is present in the resulting equation. Add an identical amount of OH^- to both sides of the equation so that all the H^+ is "neutralized," becoming water.

✔ Cancel any amounts of H_2O that appear on both sides of the equation.

That's it. Really.

Q. Balance the following redox reaction equation assuming basic conditions:

$$Cr^{2+} + Hg \rightarrow Cr + HgO$$

A. Begin balancing as if the reaction occurs under acidic conditions, and then neutralize any H^+ ions by adding OH^- equally to both sides. Finally, cancel any excess H_2O molecules.

Step 1: Divide the equation into half-reactions for oxidation and reduction by using oxidation numbers.

$$Hg \rightarrow HgO \text{ (oxidation)}$$

$$Cr^{2+} \rightarrow Cr \text{ (reduction)}$$

Step 2: Balance the half-reactions, temporarily ignoring O and H. (They're already balanced here.)

$$Hg \rightarrow HgO$$

$$Cr^{2+} \rightarrow Cr$$

Step 3: Balance the half-reactions for O and H by adding H_2O and H^+, respectively.

$$Hg + H_2O \rightarrow HgO + 2H^+$$

$$Cr^{2+} \rightarrow Cr$$

Step 4: Balance the charge within each half-reaction by adding electrons (e^-).

$$Hg + H_2O \rightarrow HgO + 2H^+ + 2e^-$$

$$Cr^{2+} + 2e^- \rightarrow Cr$$

Step 5: Balance the charge of the half-reactions with respect to each other. (They're already balanced here.)

$$Hg + H_2O \rightarrow HgO + 2H^+ + 2e^-$$

$$Cr^{2+} + 2e^- \rightarrow Cr$$

Step 6: Add the half-reactions, reuniting them within the total reaction equation.

$$Hg + Cr^{2+} + H_2O + 2e^- \rightarrow HgO + Cr + 2H^+ + 2e^-$$

Step 7: Simplify by canceling items that appear on both sides of the equation.

$Hg + Cr^{2+} + H_2O \rightarrow HgO + Cr + 2H^+$

Step 8: Neutralize H$^+$ by adding sufficient and equal amounts of OH$^-$ to both sides.

$Hg + Cr^{2+} + H_2O + 2OH^- \rightarrow HgO + Cr + 2H_2O$

Step 9: Simplify by canceling H$_2$O as possible from both sides.

$Hg + Cr^{2+} + 2OH^- \rightarrow HgO + Cr + H_2O$

---

6. Balance the following redox reaction equation assuming basic conditions:

$Cl_2 + Sb \rightarrow Cl^- + SbO^+$

Solve It

7. Balance the following redox reaction equation assuming basic conditions:

$SO_4^{2-} + ClO_2^- \rightarrow SO_2 + ClO_2$

Solve It

---

8. Balance the following redox reaction equation assuming basic conditions:

$BrO_3^- + PbO \rightarrow Br_2 + PbO_2$

Solve It

Answers to Questions on Electrons in Redox

Redox reactions have a bad reputation among chemistry students because of the perceptions that they're difficult to understand and even more difficult to balance. But the perception is only a perception. The whole process boils down to the following principles:

- ✔ Balancing redox reaction equations is exactly like balancing other equations; you simply have one more component to balance — the electron.

- ✔ Use oxidation numbers to discern the oxidation and reduction half-reactions.

- ✔ Under acidic conditions, balance O and H atoms by adding H_2O and H^+.

- ✔ Under basic conditions, balance as you do with acidic conditions, but then neutralize any H^+ by adding OH^-.

So, relax. Well . . . relax after you check your work.

1 **The oxidizing agent is MnO_2, and the reducing agent is NO_2^-.** The oxidation number of Mn is +4 in the MnO_2 reactant and is +2 in the Mn^{2+} product. The oxidation number of N is +3 in the NO_2^- reactant and is +5 in the NO_3^- product.

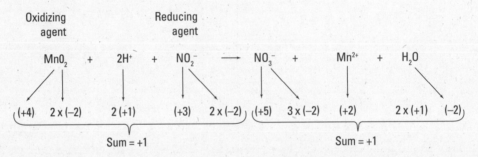

2 **The oxidizing agent is NO_3^-, and the reducing agent is H_2S.** The oxidation number of S is –2 in the H_2S reactant and is 0 in the S product. The oxidation number of N is +5 in the NO_3 reactant and is +4 in the NO_2 product.

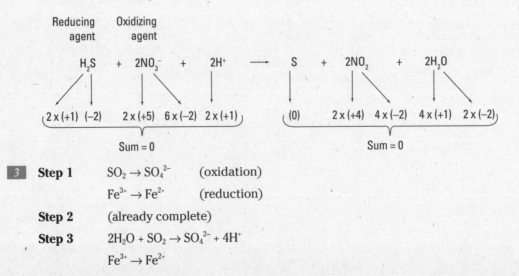

3 **Step 1** $SO_2 \rightarrow SO_4^{2-}$ (oxidation)

$Fe^{3+} \rightarrow Fe^{2+}$ (reduction)

Step 2 (already complete)

Step 3 $2H_2O + SO_2 \rightarrow SO_4^{2-} + 4H^+$

$Fe^{3+} \rightarrow Fe^{2+}$

Step 4 $2H_2O + SO_2 \rightarrow SO_4^{2-} + 4H^+ + 2e^-$

$1e^- + Fe^{3+} \rightarrow Fe^{2+}$

Step 5 $2H_2O + SO_2 \rightarrow SO_4^{2-} + 4H^+ + 2e^-$

$2e^- + 2Fe^{3+} \rightarrow 2Fe^{2+}$

Step 6 $2e^- + 2H_2O + 2Fe^{3+} + SO_2 \rightarrow 2Fe^{2+} + SO_4^{2-} + 4H^+ + 2e^-$

Step 7 $2H_2O + 2Fe^{3+} + SO_2 \rightarrow 2Fe^{2+} + SO_4^{2-} + 4H^+$

4 **Step 1** $Be \rightarrow Be_2O_3^{2-}$ (oxidation)

$Ag^+ \rightarrow Ag$ (reduction)

Step 2 $2Be \rightarrow Be_2O_3^{2-}$

$Ag^+ \rightarrow Ag$

Step 3 $3H_2O + 2Be \rightarrow Be_2O_3^{2-} + 6H^+$

$Ag^+ \rightarrow Ag$

Step 4 $3H_2O + 2Be \rightarrow Be_2O_3^{2-} + 6H^+ + 4e^-$

$1e^- + Ag^+ \rightarrow Ag$

Step 5 $3H_2O + 2Be \rightarrow Be_2O_3^{2-} + 6H^+ + 4e^-$

$4e^- + 4Ag^+ \rightarrow 4Ag$

Step 6 $4e^- + 3H_2O + 4Ag^+ + 2Be \rightarrow 4Ag + Be_2O_3^{2-} + 6H^+ + 4e^-$

Step 7 $3H_2O + 4Ag^+ + 2Be \rightarrow 4Ag + Be_2O_3^{2-} + 6H^+$

5 **Step 1** $Bi \rightarrow BiO^+$ (oxidation)

$Cr_2O_7^{2-} \rightarrow Cr^{3+}$ (reduction)

Step 2 $Bi \rightarrow BiO^+$

$Cr_2O_7^{2-} \rightarrow 2Cr^{3+}$

Step 3 $H_2O + Bi \rightarrow BiO^+ + 2H^+$

$14H^+ + Cr_2O_7^{2-} \rightarrow 2Cr^{3+} + 7H_2O$

Step 4 $H_2O + Bi \rightarrow BiO^+ + 2H^+ + 3e^-$

$6e^- + 14H^+ + Cr_2O_7^{2-} \rightarrow 2Cr^{3+} + 7H_2O$

Step 5 $2H_2O + 2Bi \rightarrow 2BiO^+ + 4H^+ + 6e^-$

$6e^- + 14H^+ + Cr_2O_7^{2-} \rightarrow 2Cr^{3+} + 7H_2O$

Step 6 $6e^- + 14H^+ + 2H_2O + Cr_2O_7^{2-} + 2Bi \rightarrow 2Cr^{3+} + 2BiO^+ + 7H_2O + 4H^+ + 6e^-$

Step 7 $10H^+ + Cr_2O_7^{2-} + 2Bi \rightarrow 2Cr^{3+} + 2BiO^+ + 5H_2O$

6 **Step 1** $Sb \rightarrow SbO^+$ (oxidation)

$Cl_2 \rightarrow Cl^-$ (reduction)

Step 2 $Sb \rightarrow SbO^+$

$Cl_2 \rightarrow 2Cl^-$

Step 3 $Sb + H_2O \rightarrow SbO^+ + 2H^+$

$Cl_2 \rightarrow 2Cl^-$

Step 4 $Sb + H_2O \rightarrow SbO^+ + 2H^+ + 3e^-$

$Cl_2 + 2e^- \rightarrow 2Cl^-$

Step 5 $2Sb + 2H_2O \rightarrow 2SbO^+ + 4H^+ + 6e^-$

$3Cl_2 + 6e^- \rightarrow 6Cl^-$

Step 6 $2Sb + 3Cl_2 + 2H_2O + 6e^- \rightarrow 2SbO^+ + 6Cl^- + 4H^+ + 6e^-$

Step 7 $2Sb + 3Cl_2 + 2H_2O \rightarrow 2SbO^+ + 6Cl^- + 4H^+$

Step 8 $2Sb + 3Cl_2 + 2H_2O + 4OH^- \rightarrow 2SbO^+ + 6Cl^- + 4H_2O$

Step 9 $2Sb + 3Cl_2 + 4OH^- \rightarrow 2SbO^+ + 6Cl^- + 2H_2O$

7 **Step 1** $ClO_2^- \rightarrow ClO_2$ (oxidation)

$SO_4^{2-} \rightarrow SO_2$ (reduction)

Step 2 (already complete)

Step 3 $ClO_2^- \rightarrow ClO_2$

$SO_4^{2-} + 4H^+ \rightarrow SO_2 + 2H_2O$

Step 4 $ClO_2^- \rightarrow ClO_2 + 1e^-$

$SO_4^{2-} + 4H^+ + 2e^- \rightarrow SO_2 + 2H_2O$

Step 5 $2ClO_2^- \rightarrow 2ClO_2 + 2e^-$

$SO_4^{2-} + 4H^+ + 2e^- \rightarrow SO_2 + 2H_2O$

Step 6 $2ClO_2^- + SO_4^{2-} + 4H^+ + 2e^- \rightarrow 2ClO_2 + SO_2 + 2H_2O + 2e^-$

Step 7 $2ClO_2^- + SO_4^{2-} + 4H^+ \rightarrow 2ClO_2 + SO_2 + 2H_2O$

Step 8 $2ClO_2^- + SO_4^{2-} + 4H_2O \rightarrow 2ClO_2 + SO_2 + 2H_2O + 4OH^-$

Step 9 $2ClO_2^- + SO_4^{2-} + 2H_2O \rightarrow 2ClO_2 + 4OH^-$

8 **Step 1** $PbO \rightarrow PbO_2$ (oxidation)

$BrO_3^- \rightarrow Br_2$ (reduction)

Step 2 $PbO \rightarrow PbO_2$

$2BrO_3^- \rightarrow Br_2$

Step 3 $PbO + H_2O \rightarrow PbO_2 + 2H^+$

$2BrO_3^- + 12H^+ \rightarrow Br_2 + 6H_2O$

Step 4 $PbO + H_2O \rightarrow PbO_2 + 2H^+ + 2e^-$

$2BrO_3^- + 12H^+ + 10e^- \rightarrow Br_2 + 6H_2O$

Step 5 $5PbO + 5H_2O \rightarrow 5PbO_2 + 10H^+ + 10e^-$

$2BrO_3^- + 12H^+ + 10e^- \rightarrow Br_2 + 6H_2O$

Step 6 $5PbO + 2BrO_3^- + 5H_2O + 12H^+ + 10e^- \rightarrow 5PbO_2 + Br_2 + 6H_2O + 10H^+ + 10e^-$

Step 7 $5PbO + 2BrO_3^- + 2H^+ \rightarrow 5PbO_2 + Br_2 + H_2O$

Step 8 $5PbO + 2BrO_3^- + 2H_2O \rightarrow 5PbO_2 + Br_2 + H_2O + 2OH^-$

Step 9 $5PbO + 2BrO_3^- + H_2O \rightarrow 5PbO_2 + Br_2 + 2OH^-$

Chapter 19

Galvanizing Yourself into Electrochemistry

Although we're sure that you're thrilled to hear that redox reactions are the driving force behind the creation of rust and the greening of the copper domes on cathedrals, you should know that you don't need to sit around watching metal rust to see redox reactions in action. In fact, they play a very important role in your everyday life. The redox reactions that you find out about in Chapter 18 are the silent partner backing the unstoppable Energizer Bunny. That's right! Redox reactions are the essential chemistry behind the inner workings of the all-important battery. In fact, an entire branch of chemistry, called *electrochemistry*, centers around the study of electrochemical cells, which create electrical energy from chemical energy. We introduce the basics in this chapter.

Identifying Anodes and Cathodes

The energy provided by batteries is created in a unit called a *voltaic* or *galvanic cell.* Many batteries use a number of voltaic cells wired in series, and others use a single cell. Voltaic cells harness the energy released in a redox reaction and transform it into electrical energy. A voltaic cell is created by connecting two metals called *electrodes* in solution with an external circuit. In this way, the reactants aren't in direct contact, but can transfer electrons to one another through an external pathway, fueling the redox reaction.

The electrode that undergoes oxidation is called the *anode.* You can easily remember this if you burn the phrase "an ox" (for *an*ode *ox*idation) into your memory. The phrase "red cat" is equally useful for remembering what happens at the other electrode, called the *cathode,* where the reduction reaction occurs.

Electrons created in the oxidation reaction at the anode of a voltaic cell flow along an external circuit to the cathode, where they fuel the reduction reaction taking place there. We use the spontaneous reaction between zinc and copper as an example of a voltaic cell here, but you should realize that many powerful redox reactions power many types of batteries, so they're not limited to reactions between copper and zinc.

Zinc metal reacts spontaneously with an aqueous solution of copper sulfate when they're placed in direct contact with one another. Zinc, being a more reactive metal than copper (it's higher on the activity series of metals presented in Chapter 8), displaces the copper ions in solution. The displaced copper deposits itself as pure copper metal on the surface of the dissolving zinc strip. At first, it may appear to be a simple single replacement reaction, but it's also a redox reaction.

The oxidation of zinc proceeds as $Zn(s) \rightarrow Zn^{2+}(aq) + 2e^-$. The two electrons created in this oxidation of zinc are consumed by the copper in the reduction half of the reaction $Cu^{2+}(aq) + 2e^- \rightarrow Cu(s)$. This makes the total reaction $Cu^{2+}(aq) + 2e^- + Zn(s) \rightarrow Cu(s) + Zn^{2+}(aq) + 2e^-$. The electron duo appears on both sides of this combined reaction, revealing its identity as a spectator, so the reaction is really just $Cu^{2+}(aq) + Zn(s) \rightarrow Cu(s) + Zn^{2+}(aq)$.

This reaction takes place when the two are in direct contact, but as we explain earlier in this section, a voltaic cell is created by connecting the two reactants by an external pathway. Only the electrons created at the anode in the oxidation reaction can travel to the reduction half of the reaction along this external pathway. A voltaic cell using this same oxidation-reduction reaction between copper and zinc is shown in Figure 19-1, which we examine piece by piece.

✔ First, note that zinc (Zn) is being oxidized at the anode, which is labeled with a negative sign. This doesn't mean that the anode is negatively charged. Rather, it means that electrons are being created there. The oxidation of zinc releases Zn^{2+} cations into the solution as well as two electrons that flow along the circuit to the cathode. This oxidation thus results in an increase of Zn^{2+} ions into the solution and a decrease in the mass of the zinc metal anode.

✔ The electrons created by the oxidation of zinc at the anode fuel the reduction of copper (Cu) at the cathode. This pulls Cu^{2+} from solution and deposits more Cu metal on the cathode. The result is the exact opposite of the effect occurring at the anode: The solution becomes less concentrated as Cu^{2+} ions are used up in the reduction reaction, and the electrode gains mass as Cu metal is deposited.

✔ No doubt you noticed that Figure 19-1 also contains a U-shaped tube connecting the two solutions. This is called the *salt bridge,* and it serves to correct the charge imbalance created as the anode releases more and more cations into its solution, resulting in a net positive charge, and the cathode uses up more and more of the cations in its solution, leaving it with a net negative charge. The salt bridge contains an electrolytic salt (in this case $NaNO_3$). A good salt bridge is created with an electrolyte whose component ions won't react with the ions already in solution. The salt bridge functions by sucking up the extra NO_3^- ions in the cathode solution and depositing NO_3^- ions from the other end of the bridge into the anode solution. It also sucks up the excess positive charge at the anode by absorbing Zn^{2+} ions and depositing its own cation Na^+ into the cation solution. This is necessary because the solutions must be neutral for the redox reaction to continue. It also completes the cell's circuit by allowing for the flow of charge back to the anode.

A useful way to remember how ions flow in the salt bridge is that anions travel to the anode, while cations travel to the cathode.

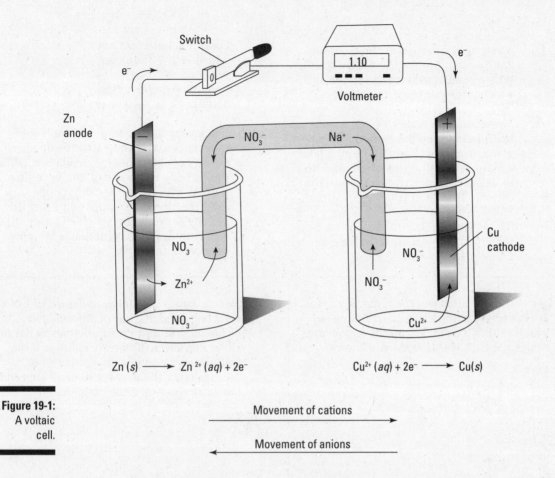

Switch

1.10

Voltmeter

e^-

e^-

Zn
anode

−

+

NO_3^-

Na^+

NO_3^-

Zn^{2+}

NO_3^-

NO_3^-

NO_3^-

Cu^{2+}

Cu
cathode

$Zn(s) \longrightarrow Zn^{2+}(aq) + 2e^-$

$Cu^{2+}(aq) + 2e^- \longrightarrow Cu(s)$

Figure 19-1:
A voltaic
cell.

Movement of cations

Movement of anions

These voltaic cells can't run forever, however. The loss of mass at the zinc anode will eventually exhaust the supply of zinc, and the redox reaction won't be able to continue. This phenomenon is why most batteries run out over time. Rechargeable batteries take advantage of a reverse reaction to resupply the anode, but many redox reactions don't allow for this, so rechargeable batteries must be made of very specific reactants.

The voltage provided by a battery depends largely on the materials that make up the two electrodes. The reactions carried out in the 1.5V batteries that power your calculator, flashlight, or MP3 player, for example, are often reactions between carbon and zinc-chloride or zinc and manganese-dioxide. The high-voltage, long-lived batteries that power your computer, pacemaker, or watch, on the other hand, generally have anodes composed of a lithium compound, which can provide roughly twice the voltage of many other batteries and are longer-lived (certainly a desirable quality in the battery that powers a pacemaker). Although many batteries contain a single voltaic cell, a number of battery types that require high voltages, such as car batteries, harness the power of multiple voltaic cells by wiring them in series with one another. A 12V car battery, for example, is created by wiring six 2V voltaic cells together.

Voltage is measured by attaching a voltmeter to a circuit as shown in Figure 19-1. When the switch is closed, the voltmeter reads the potential difference between the anode and cathode of the cell.

Q. A voltaic cell harnesses the reaction $2Al(s) + 3Sn^{2+}(aq) \rightarrow 2Al^{3+}(aq) + 3Sn(s)$. Which metal makes up the anode and which makes up the cathode?

A. **The aluminum makes up the anode, and the tin makes up the cathode.** The best place to start is to add in the spectator electrons to both sides of the equation to balance the positive charges of the cations. This gives you the equation $2Al(s) + 3Sn^{2+}(aq) + 6e^- \rightarrow 2Al^{3+}(aq) + 6e^- + 3Sn(s)$. Next, isolate the oxidation and reduction half-reactions (as we explain in Chapter 18):

Oxidation half-reaction:
$2Al(s) \rightarrow 2Al^{3+}(aq) + 6e^-$

Reduction half-reaction:
$3Sn^{2+}(aq) + 6e^- \rightarrow 3Sn(s)$

Using your "an ox" and "red cat" mnemonics, you know that the anode is the site of aluminum (Al) oxidation, while the cathode is the site of tin (Sn) reduction. As with the earlier zinc/copper example, this is also apparent using the activity series of metals (see Chapter 8), which shows that aluminum is far more reactive than tin.

1. Write the oxidation and the reduction halves of a reaction between cadmium (II) and tin (II). Which makes up the anode and which the cathode? How do you know?

Solve It

2. Sketch a diagram similar to Figure 19-1 of a cell composed of a bar of chromium in a chromium (III) nitrate solution and a bar of silver in a silver (II) nitrate solution if the salt bridge is composed of potassium nitrate. Trace the movement of all ions and label the anode and the cathode.

Solve It

Calculating Electromotive Force and Standard Reduction Potentials

In the previous section, we take for granted that charge can flow through the external circuit connecting the two halves of a voltaic cell, but what actually causes this flow of charge? The answer lies in a concept called *potential energy*. When a difference in potential energy is established between two locations, an object has a natural tendency to move from an area of higher potential energy to an area of lower potential energy.

When Wile E. Coyote places a large circular boulder at the top of a hill overlooking a road on which he knows his nemesis the Road Runner will soon be traveling, he takes it for granted that when he releases the boulder it will simply roll down the hill. This is due to the difference in potential energy between the top and bottom of the hill, which have high and low potential energies respectively (although Coyote will need to be far more well versed in physics if he's to time the crushing of his adversary properly).

In a similar, though less diabolical, manner, the electrons produced at the anode of a voltaic cell have a natural tendency to flow along the circuit to a location with lower potential: the cathode. This potential difference between the two electrodes causes the *electromotive force*, or *EMF*, of the cell. EMF is also often referred to as the *cell potential* and is denoted E_{cell}. The cell potential varies with temperature and concentration of products and reactants and is measured in *volts (V)*. The E_{cell} that occurs when concentrations of solutions are all at $1M$ and cell is at standard temperature and pressure (STP) is given the special name of *standard cell potential*, or E^0_{cell}.

Much like assigning enthalpies to pieces of a reaction and summing them using Hess's Law (see Chapter 15), cell potentials can be tabulated by taking the difference between the standard potentials of the oxidation and reduction half-reactions separately. To do this properly, chemists had to choose either the oxidation or reduction half-reactions to be positive and the other to be negative. They happened to assign positive potentials to reduction half-reactions and negative potentials to oxidation half-reactions. The standard potential at an electrode is therefore a measurement of its propensity to undergo a reduction reaction. As such, these potentials are often referred to as *standard reduction potentials*. They are calculated using the formula

$$E^0_{cell} = E^0_{red}(\text{cathode}) - E^0_{red}(\text{anode})$$

Table 19-1 lists some standard reduction potentials along with the reduction half-reactions associated with them. The table is ordered from the most negative E^0_{cell} (most likely to oxidize) to the most positive E^0_{cell} (most likely to be reduced). The reactions with negative E^0_{cell} are therefore reactions that happen at the anode of a voltaic cell, while those with a positive E^0_{cell} occur at the cathode.

Table 19-1 Reduction Half-Reactions and Standard Reduction Potentials	
Reduction Half-Reaction	E^0_{cell} *(in V)*
$Li^+(aq) + e^- \rightarrow Li(s)$	−3.04
$K^+(aq) + e^- \rightarrow K(s)$	−2.92
$Ca^{2+}(aq) + 2e^- \rightarrow Ca(s)$	−2.76
$Na^+(aq) + e^- \rightarrow Na(s)$	−2.71
$Zn^{2+}(aq) + 2e^- \rightarrow Zn(s)$	−0.76
$Fe^{2+}(aq) + 2e^- \rightarrow Fe(s)$	−0.44
$2H^+(aq) + 2e^- \rightarrow H_2(g)$	0
$Cu^{2+}(aq) + 2e^- \rightarrow Cu(s)$	+0.34
$I_2(s) + 2e^- \rightarrow 2I^-(aq)$	+0.54
$Ag^+(aq) + e^- \rightarrow Ag(s)$	+0.80

(continued)

Table 19-1 (continued)

Reduction Half-Reaction	E^0_{cell} (in V)
$Br_2(l) + 2e^- \rightarrow 2Br^-(aq)$	+1.06
$O_3(g) + 2H^+(aq) + 2e^- \rightarrow O_2(g) + H_2O(l)$	+2.07
$F_2(g) + 2e^- \rightarrow 2F^-(aq)$	+2.87

Note that not all the reactions in Table 19-1 show the oxidation or reduction of solid metals as in our examples so far. Liquids and gases are thrown in as well. Not every voltaic cell is fueled by a reaction taking place between the metals of the electrodes. Although the cathode itself must be made of a metal to allow for the flow of electrons, those electrons can be passed into a gas or a liquid to complete the reduction half-reaction. Examine Figure 19-2 for an example of such a cell, which includes a gaseous electrode.

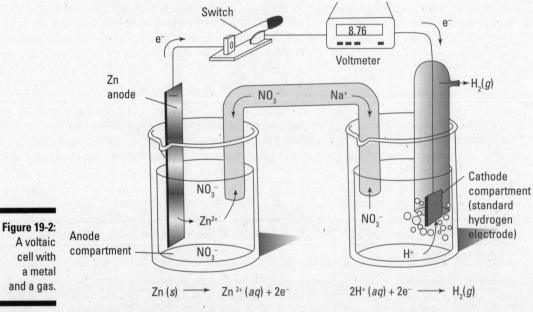

Figure 19-2:
A voltaic cell with a metal and a gas.

$$Zn(s) \longrightarrow Zn^{2+}(aq) + 2e^-$$

$$2H^+(aq) + 2e^- \longrightarrow H_2(g)$$

An equation relates standard cell potential to the EMF of the cell. This equation, called the *Nernst equation,* is expressed as

$$E_{cell} = E^0_{cell} - (RT / nF) \ln Q$$

Where Q is the reaction quotient (discussed in Chapter 14), n is the number of electrons transferred in the redox reaction, R is the universal gas constant (8.21J/(K × mol)), T is the temperature in Kelvin, and F is the Faraday constant (9.65×10^4 coulombs/mol, where coulombs are a unit of electric charge). Recall also that ln stands for "natural log" and is a function in your calculator. With this information, you can assign quantitative values to the EMFs of batteries. The equation also reveals that the EMF of a battery depends on temperature, which is why batteries are less likely to function well in the cold.

Q. What is the EMF of the cell shown in Figure 19-1 when it is at a temperature of 25°C if Cu^{2+} is 0.1*M* and Zn^{2+} is 3.0*M*?

A. **1.06V.** To find the EMF, or E_{cell}, you first need to determine the standard cell potential. Do this by looking up the oxidation and reduction half-reactions in Table 19-1. The oxidation of zinc has an E^0_{cell} of –0.76, while the reduction of copper has an E^0_{cell} of 0.34. Recognizing that copper is the cathode and zinc the anode, you can plug these values into the equation $E^0_{cell} = E^0_{red}(\text{cathode}) - E^0_{red}(\text{anode})$ to get $E^0_{cell} = 0.34V + 0.76V = 1.10V$.

You must then use the Nernst equation to calculate the EMF, which means you must determine the values of Q and n. You can find n by examining the oxidation and reduction half-reactions presented earlier in this chapter, which indicate that two electrons are exchanged in the process. Q is expressed as $Q = \dfrac{\left[Zn^{2+}\right]}{\left[Cu^{2+}\right]} = \dfrac{\left[3.0\right]}{\left[0.1\right]} = 30$ (if you can't recall how to calculate reaction quotients, see Chapter 14). Next, write out the Nernst equation for this specific cell.

$$E_{cell} = E^0_{cell} - \left(\frac{(8.21J/K \times mol) \times T}{n \times (9.65 \times 10^4 \, Coulombs/mol)}\right) \times \ln\left(\frac{\left[Zn^{2+}\right]}{\left[Cu^{2+}\right]}\right)$$

Plugging in known values for E^0_{cell}, n, Cu^{2+}, and Zn^{2+}, and converting temperature to degrees Kelvin, you get

$$E_{cell} = 1.10V - \left(\frac{(8.21J/K \times mol) \times 298K}{2 \times (9.65 \times 10^4 \, Coulombs/mol)}\right) \times \ln(30) = 1.06V$$

3. Using the two half-reactions fueling the cell in Figure 19-2, calculate the EMF of that cell at a temperature of 25°C and at 0°C given concentrations of 10M and 0.01M for H$^+$ and Zn^{2+} respectively.

Solve It

4. Is EMF constant over the life of a voltaic cell? Why or why not?

Solve It

Coupling Current to Chemistry: Electrolytic Cells

A positive standard cell potential tells you that the cathode is at a higher potential than the anode, and the reaction is therefore spontaneous. What do you do with a cell that has a negative E^0_{cell}? Electrochemical cells that rely on such nonspontaneous reactions are called *electrolytic cells*. The redox reactions in electrolytic cells rely on a process called *electrolysis*. These reactions require that a current be passed through the solution, forcing it to split into components that then fuel the redox reaction. Such cells are created by applying a current source, such as a battery, to electrodes placed in a solution of *molten salt*, or salt which has been heated until it melts. This splits the ions that make up the salt.

Cations have a natural tendency to migrate toward the negative anode, where they're oxidized, and anions migrate toward the positive cathode and are reduced. Thus the "an ox" and "red cat" mnemonics are still valid.

Figure 19-3 shows an electrolytic cell using molten sodium chloride. A redox reaction between sodium and chlorine won't happen spontaneously, but the electrical energy produced by the battery provides the additional energy needed to fuel the reaction. In the process, chlorine anions are oxidized at the anode, creating chlorine gas, and sodium is reduced at the cathode and is deposited onto it as sodium metal.

The amount of pure metal created in an electrolytic cell can be analyzed quantitatively. First, you need to calculate the number of electrons that must be created to fuel electrolysis. To do this, write out the reduction half-reaction and determine how many electrons are needed to accomplish it. In the cell shown in Figure 19-3, reduction of sodium at the cathode occurs according to the equation $Na^+(aq) + e^- \rightarrow Na(s)$, thus 1 mole of electrons can reduce aqueous sodium to produce 1 mole of sodium metal. If the cathode were reducing copper metal according to the reaction $Cu^{2+}(aq) + 2e^- \rightarrow Cu(s)$, it would require 2 moles of electrons for every mole of copper created, and so on.

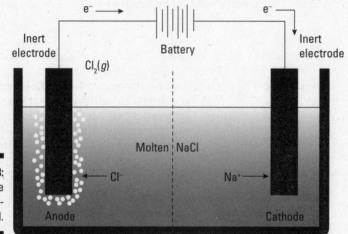

To determine how many moles of metal have been deposited, you need to determine how many electrons have flowed through the circuit to fuel this reaction. You can determine the amount of charge passing through the circuit (measured in Coulombs, the standard unit of charge) by using the current provided by the power source and the amount of time the cell operates. This is done using the relationship

Charge = Current in amperes × Time in seconds

This works out because the *ampere* (the standard unit of current; abbreviated A) is defined as 1 coulomb per second. Because this gives us the amount of charge that has passed through the circuit during its operating time, all that remains is to calculate the number of moles of electrons that make up that amount of charge. For this, you use the conversion factor 1 mol e⁻ = 96,500 coulombs.

EXAMPLE

Q. How many grams of sodium metal would be deposited onto the cathode of the electrolytic cell shown in Figure 19-3 if it was connected to a battery providing a 15A current for 1.5 hours?

A. **19.32g Na(s).** Start by determining the amount of charge created in the cell. To do this, you must multiply the current by the time in seconds, which requires the conversion

$$\frac{1.5 \text{ hr}}{1} \times \frac{60 \text{ min}}{\text{hr}} \times \frac{60 \text{ sec}}{\text{min}} = 5.4 \times 10^3 \text{ sec}$$

Plugging this value for time into the equation Charge = Current in amperes × Time in seconds yields 8.1×10^4 coulombs of charge. Convert this to moles of electrons.

$$\frac{8.1 \times 10^4 \text{ Coulombs}}{1} \times \frac{1 \text{ mol e}^-}{96,500 \text{ Coulombs}} = 0.84 \text{ mol e}^-$$

Because you know that 1 mole of electrons can create 1 mole of sodium metal, this translates to 0.84 mol Na(s) produced. All that remains is to convert this to grams using the gram atomic mass (see Chapter 7 for details).

$$\frac{0.84 \text{ mol Na}(s)}{1} \times \frac{23.0 \text{g Na}}{1 \text{ mol Na}} = 19.32 \text{g Na}(s)$$

5. How much sodium would be deposited in the same amount of time as in the example question if the battery could supply a 50A current?

Solve It

6. If an electrolytic cell containing molten aluminum (III) chloride were attached to a battery providing 8.0A for 45 minutes, how many grams of pure aluminum metal would be deposited on the cathode in that time?

Solve It

Answers to Questions on Electrochemistry

Are you charged up to check your answers? Take a gander at this section to see how you did on the practice problems in this chapter.

1 You actually need to evaluate the second portion of this question first, which can be answered with a simple glance at the activity series of metals in Chapter 8, which tells you that cadmium is more active than tin. So it must be cadmium that is oxidized and makes up the anode, and tin that is reduced and makes up the cathode. The half-reactions are therefore

Oxidation half-reaction: $Cd(s) \rightarrow Cd^{2+}(aq) + 2e^-$

Reduction half-reaction: $Sn^{2+}(aq) + 2e^- \rightarrow Sn(s)$

2 Chromium is more active than silver (as we explain in Chapter 8), so chromium is oxidized in this reaction and silver reduced according to the half-reactions

Oxidation half-reaction: $Cr(s) \rightarrow Cr^{3+}(aq) + 3e^-$

Reduction half-reaction: $Ag^{2+}(aq) + 2e^- \rightarrow Ag(s)$

The balanced redox reaction is therefore $2Cr(s) + 3Ag^{2+}(aq) + 6e^- \rightarrow 2Cr^{3+}(aq) + 6e^- + 3Ag(s)$; see Chapter 18 for details on balancing redox reactions.

Your diagram should therefore look like the following figure:

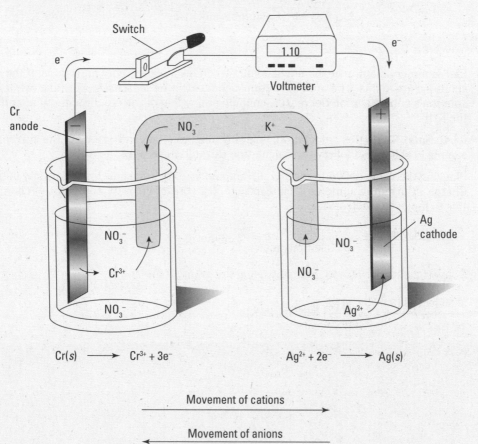

3 **0.848V for 25°C and 0.840V for 0°C.**

Oxidation half-reaction: $Zn(s) \rightarrow Zn^{2+}(aq) + 2e^-$

Reduction half-reaction: $2H^+(aq) + 2e^- \rightarrow H_2(g)$

The potential for the oxidation half-reaction is revealed in Table 19-1 to be –0.76V, while the reduction half is shown to be 0V. Plugging these values into the equation $E^0_{cell} = E^0_{red}(cathode) - E^0_{red}(anode)$ gives you $E^0_{cell} = 0V + 0.76V = 0.76V$. The problem asks you to calculate the E_{cell} at T = 298K and at T = 273K and tells you that $H^+ = 10M$ and $Zn^{2+} = 0.01M$. The reaction quotient Q is

$$Q = \frac{\left[Zn^{2+}\right]}{\left[H^+\right]^2} = \frac{0.01}{10} = 0.001$$

Plugging all of these values into the Nernst equation yields

$$E_{cell} = 0.76V - \left(\frac{(8.21J/K \times mol) \times 298K}{2 \times (9.65 \times 10^4 Coulombs/mol)}\right) \times \ln(0.001) = 0.848V \quad \text{for } 25°C$$

$$E_{cell} = 0.76V - \left(\frac{(8.21J/K \times mol) \times 273K}{2 \times (9.65 \times 10^4 Coulombs/mol)}\right) \times \ln(0.001) = 0.840V \quad \text{for } 0°C$$

4 EMF is not constant over the life of a voltaic cell because the concentrations of the aqueous solutions are in flux. The anode solution's concentration increases over time, while the cathode solution's concentration decreases, changing the value of the reaction quotient and therefore the EMF.

5 **64.4g Na(s).** To do this calculation, you only need to change the value of the current in the sample problem and follow the calculation through the rest of the way.

Plugging the value of 50A in for the current and keeping the same value for time in the equation Charge = Current in amperes × Time in seconds, you get 2.7×10^5 coulombs of charge. Convert this to moles of electrons.

$$\frac{2.7 \times 10^5 \text{ Coulombs}}{1} \times \frac{1 \text{ mol } e^-}{96,500 \text{ Coulombs}} = 2.80 \text{ mol } e^-$$

Convert 2.80 mol Na(s) to grams by using the gram atomic mass.

$$\frac{2.80 \text{ mol Na}(s)}{1} \times \frac{23.0g \text{ Na}}{1 \text{ mol Na}} = 64.4g \text{ Na}(s)$$

6   **1.89g Al(*s*).** First, you have to write out the reduction half of the reaction to determine how many moles of electrons you need to reduce the aluminum.

Reduction half-reaction: $Al^{3+}(aq) + 3e^- \rightarrow Al(g)$

This means that 3 moles of electrons are required to reduce each mole of aluminum. Next convert 45 minutes into seconds, which gives you 2,700 sec. Multiply this by the current of 8 amperes to get 21,600 coulombs. Divide this by 96,500 to give you 0.22 mol of electrons produced. Divide this number by 3 to give you 0.07 mol aluminum. Lastly, multiply the number of moles of aluminum by its gram atomic mass (which is 27 grams) to give you 1.89g Al(*s*).

Chapter 20

Doing Chemistry with Atomic Nuclei

· ·

In This Chapter

▶ Decaying into alpha, beta, and gamma

▶ Living with half-lives

▶ Combining and splitting nuclei with fusion and fission

· ·

As described in Chapter 3, many elements in the periodic table exist in unstable versions called radioisotopes. These radioisotopes decay into other (usually more stable) elements in a process called radioactive decay. Because the stability of these radioisotopes depends on the composition of their nuclei, radioactivity is considered a form of *nuclear chemistry*. Unsurprisingly, nuclear chemistry deals with nuclei and nuclear processes. Nuclear fusion, which fuels our sun, and nuclear fission, which fuels a nuclear bomb, are examples of nuclear chemistry because they deal with the joining or splitting of atomic nuclei. In this chapter, you find out about nuclear decay, rates of decay called half-lives, and the processes of fusion and fission.

Decaying Nuclei in Different Ways

Many radioisotopes exist, but not all radioisotopes are created equal. Radioisotopes break down through three separate decay processes (or decay *modes*): alpha decay, beta decay, and gamma decay.

Alpha decay

The first type of decay process, called *alpha decay*, involves emission of an alpha particle by the nucleus of an unstable atom. An alpha particle (α particle) is nothing more exotic than the nucleus of a helium atom, made of two protons and two neutrons. Emitting an alpha particle decreases the atomic number of the daughter nucleus by 2 and decreases the mass number by 4. (See Chapter 3 for details about atomic numbers and mass numbers.) In general, then, a parent nucleus X decays into a daughter nucleus Y and an alpha particle.

$$_B^A X \rightarrow _{B-2}^{A-4} Y + _2^4 He$$

Beta decay

The second type of decay, called *beta decay* (β decay), comes in three forms, termed *beta-plus*, *beta-minus*, and *electron capture*. All three involve emission or capture of an electron or a *positron* (a particle with the tiny mass of an electron but with a positive charge), and all three also change the atomic number of the daughter atom.

✔ **In beta-plus decay, a proton in the nucleus decays into a neutron, a positron (e^+), and a tiny, weakly interacting particle called a** *neutrino* **(ν).** This decay decreases the atomic number by 1. The mass number, however, *does not change*. Both protons and neutrons are *nucleons* (particles in the nucleus), after all, each contributing 1 atomic mass unit. The general pattern of beta-plus decay is shown here:

$$_B^A X \rightarrow \, _{B-1}^A Y + e^+ + \upsilon$$

✔ **Beta-minus decay essentially mirrors beta-plus decay: A neutron converts into a proton, emitting an electron and an** *anti***neutrino (which has the same symbol as a neutrino except for the line on top).** Particle and antiparticle pairs such as neutrinos and antineutrinos are a complicated physics topic, so we'll keep it basic here by saying that a neutrino and an antineutrino would annihilate one another if they ever touched, but are otherwise very similar. Again, the mass number remains the same after decay because the number of nucleons remains the same. However, the atomic number *increases* by 1 because the number of protons increases by 1:

$$_B^A X \rightarrow \, _{B+1}^A Y + e^- + \bar{\upsilon}$$

✔ **The final form of beta decay, electron capture, occurs when an inner electron (one in an orbital closest to the atomic nucleus; see Chapter 4 for more about orbitals) is "captured" by an atomic proton.** By capturing the electron, the proton converts into a neutron and emits a neutrino. Here again, the atomic number decreases by 1:

$$_B^A X + e^- \rightarrow \, _{B-1}^A Y + \upsilon$$

Gamma decay

The final form of decay, termed *gamma decay* (γ decay), involves emission of a *gamma ray* (a high-energy form of light) by an excited nucleus. Although gamma decay changes neither the atomic number nor the mass number of the daughter nucleus, it often accompanies alpha or beta decay. Gamma decay allows the nucleus of a daughter atom to reach its lowest possible energy (most favorable) state. The general form of gamma decay is shown here, where $_B^A X^*$ represents the excited state of the parent nucleus, and the Greek letter gamma (γ) represents the gamma ray.

$$_B^A X^* \rightarrow \, _B^A Y + \gamma$$

EXAMPLE

Q. If a parent nucleus has decayed into $^{218}_{86}$Rn through alpha decay, what was the parent nucleus and what other particles were produced?

A. **The parent nucleus was radium, and an alpha particle was produced.** Alpha decay lowers the mass number by 4 and the atomic number by 2, so the parent nucleus must have had a mass number of 222 and an atomic number of 88. Consulting your periodic table (check out an example in Chapter 4), you find that an atom with an atomic number of 88 is radium, so the parent nucleus must have been $^{222}_{88}$Ra .

---

1. Write out the complete formula for the alpha decay of a uranium nucleus with 238 nucleons.

Solve It

2. Sodium-22, a radioisotope of sodium with 22 nucleons, decays through electron capture. Write out the complete formula below, including all emitted particles.

Solve It

---

3. Classify the following reactions as alpha, beta, or gamma decay (specify the sub-type if it's beta decay), and supply the missing particles.

a. $^{137}_{55}$Cs → ___ + e^- + ___   Type:

b. $^{241}_{95}$Am → ___ + $^4_2$He   Type:

c. $^{60}_{28}$Ni* → ___ + ___   Type:

d. $^{11}_6$C → ___ + ___ + υ Type:

Measuring Rates of Decay: Half-Lives

The word "radioactive" sounds scary, but science and medicine are stuffed with useful, friendly applications for radioisotopes. Many of these applications are centered on the predictable decay rates of various radioisotopes. These predictable rates are characterized by *half-lives.* The half-life of a radioisotope is simply the amount of time it takes for exactly half of a sample of that isotope to decay into daughter nuclei. For example, if a scientist knows that a sample originally contained 42 mg of a certain radioisotope and measures 21 mg of that isotope in the sample four days later, then the half-life of that radioisotope is four days. The half-lives of radioisotopes range from seconds to billions of years.

Radioactive dating is the process scientists use to date samples based on the amount of radioisotope remaining. The most famous form of radioactive dating is carbon-14 dating, which has been used to date human remains and other organic artifacts. However, radioisotopes have also been used to date the earth, the solar system, and even the universe. (Flip back to Chapter 3 to find out more about carbon-14 dating.)

Listed in Table 20-1 are some of the more useful radioisotopes, along with their half-lives and decay modes (we discuss these modes earlier in this chapter).

Table 20-1	Common Radioisotopes, Half-Lives, and Decay Modes	
Radioisotope	*Half-Life*	*Decay Mode*
Carbon-14	5.73×10^3 years	beta
Iodine-131	8.0 days	beta, gamma
Potassium-40	1.25×10^9 years	beta, gamma
Radon-222	3.8 days	alpha
Thorium-234	24.1 days	beta, gamma
Uranium-238	4.46×10^9 years	alpha

To calculate the remaining amount of a radioisotope, use the following formula:

$$A = A_o \times (0.5)^{t/T}$$

where A_0 is the amount of the radioisotope that existed originally, t is the amount of time the sample has had to decay, and T is the half-life.

Q. If a sample originally contained 1g of thorium-234, how much of that isotope will the sample contain one year later?

A. **2.76x10⁻⁵g.** Table 20-1 tells us that thorium-234 has a half-life of 24.1 days, so that's *T*. The time elapsed (*t*) is 365 days, and the original sample was 1g (*A₀*). Plugging these values into the half-life equation gives:

$$A = 1g \times (0.5)^{365\,days/24.1\,days} = 2.76 \times 10^{-5}g$$

Notice that the units in the exponent cancel out, so you're left with units of grams in the end. These units make sense because you're measuring the mass of sample remaining.

4. If you start with 0.5g of potassium-40, how long will it take for the sample to decay to 0.1g?

5. If a 50g sample of a radioactive element has decayed into 44.3g after 1,000 years, what is the element's half-life? Based on Table 20-1, which radioisotope do you think you're dealing with?

Making and Breaking Nuclei: Fusion and Fission

Fission and fusion differ from radioactive decay in that they require the nucleus of a parent atom to interact with an outside particle. Because the forces which hold together atomic nuclei are ridiculously powerful, the energy involved in splitting or joining two nuclei is tremendous. Here are the differences between fusion and fission:

✔ In nature, nuclear reactions occur only in the very center of stars like our sun, where extreme temperatures cause atoms of hydrogen, helium, and other light elements to smash together and join into one. This extremely energetic process, called *nuclear fusion,* is ultimately what causes the sun to shine. Fusion also provides the outward pressure required to support the sun against gravitational collapse, an event that is even more dramatic than it sounds.

> ✔ *Nuclear fission,* the splitting of an atomic nucleus, doesn't occur in nature. Humans first harnessed the tremendous power of fission during the Manhattan Project, an intense, hush-hush effort by the United States that led to the development of the first atomic bomb in 1945. Fission has since been used for more benign purposes in nuclear power plants. Nuclear power plants use a highly regulated process of fission to produce energy much more efficiently than is done in traditional, fossil fuel–burning power plants.

Fission and fusion reactions can be distinguished from one another with a simple glance at products and reactants. If the reaction shows one large nucleus splitting into two smaller nuclei, then it's most certainly fission, whereas a reaction showing two small nuclei combining to make a single heavier nucleus is definitely fusion.

Q. What type of nuclear reaction would you expect plutonium-239 to undergo and why?

A. Fission is the expected reaction. Plutonium has an exceptionally large number of nucleons and is therefore likely to undergo fission. Fusion is impossible in all elements heavier than iron, and plutonium is much much heavier than iron.

6. What type of nuclear reaction is shown in the following equation? How do you know? Where might such a reaction take place?

$$^{235}_{92}U + ^{1}_{0}n \rightarrow ^{142}_{56}Ba + ^{91}_{36}Kr + 3^{1}_{0}n$$

7. What type of nuclear reaction is shown in the following equation? How do you know? Where might such a reaction take place? There is something atypical about the two hydrogen reactants. What is it?

$$^{3}_{1}H + ^{2}_{1}H \rightarrow ^{4}_{2}He + ^{1}_{0}n$$

Solve It

Answers to Questions on Nuclear Chemistry

The following are the answers to the practice problems presented in this chapter.

1 $^{238}_{92}U \rightarrow {}^{234}_{90}Th + {}^{4}_{2}He$

Alpha decay results in the emission of a helium nucleus from the parent atom, which leaves the daughter atom with two fewer protons and a total of four fewer nucleons. In other words, the atomic number of the daughter atom is 2 fewer than the parent atom, and the mass number is 4 fewer. Because the proton number defines the identity of an atom, the element with two fewer protons than uranium must be thorium.

2 $^{22}_{11}Na + e^- \rightarrow {}^{22}_{10}Ne + \upsilon$

Electron capture is a form of beta decay that results in the atomic number decreasing by 1 and the mass number remaining the same.

3 **a.** $^{137}_{55}Cs \rightarrow {}^{137}_{56}Ba + e^- + \bar{\upsilon}$ Type: **Beta-minus**

    b. $^{241}_{95}Am \rightarrow {}^{237}_{93}Np + {}^{4}_{2}He$ Type: **Alpha**

    c. $^{60}_{28}Ni^* \rightarrow {}^{60}_{28}Ni + \gamma$ Type: **Gamma**

    d. $^{11}_{6}C \rightarrow {}^{11}_{5}B + e^+ + \upsilon$ Type: **Beta-plus**

4 **2.9×10^9 years.** The problem tells you that $A = 0.1$ and $A_0 = 0.5$. Table 20-1 tells you that the half-life of potassium-40 is 1.25×10^9 years. Plugging these values into the equation, you get

$$0.1 = 0.5 \times (0.5)^{t/1.25 \times 10^9 \text{ years}}$$

Divide both sides by 0.5 to get

$$0.2 = (0.5)^{t/1.25 \times 10^9 \text{ years}}$$

Take the natural log of both sides to get

$$\ln(0.2) = \ln\left((0.5)^{t/1.25 \cdot 10^9 \text{ years}}\right)$$

This step allows you to isolate the exponent by pulling it out in front on the right-hand side, giving

$$\ln(0.2) = \frac{t}{1.25 \times 10^9 \text{ years}} \times \ln(0.5)$$

Compute the two natural logs to get

$$-1.61 = \frac{t}{1.25 \times 10^9 \,\text{years}} \times -0.69$$

Divide both sides by –0.69 to get

$$2.32 = \frac{t}{1.25 \times 10^9}$$

Then multiply by 1.25×10^9 to get $t = 2.9 \times 10^9$ years.

5 **5,726 years; carbon-14.** The problem tells us that $A_0 = 50$g, $A = 44.3$g, and $t = 1,000$. Use the same process as in Question 4 to isolate the exponent, this time solving for T instead of t. Doing so gives us 5,726 years, which closely matches the half-life of carbon-14 in Table 20-1. (Try to keep the numbers in your calculator throughout each step of the process, and don't round until the end of the problem.)

6 This reaction is a fission reaction. It shows a heavy uranium nucleus being bombarded by a neutron and decaying into two lighter nuclei (barium and krypton). This is the very reaction that takes place in a nuclear reactor.

7 This reaction is a fusion reaction. It shows two light nuclei combining to form one heavy nucleus. This reaction fuels our sun. The two hydrogen reactants are atypical because they are unusual isotopes of hydrogen, called tritium and deuterium, respectively.

Part V
Going Organic

"I love this time of year when the organic chemistry students start exploring new and exciting ways for bonding carbon atoms."

In this part . . .

Carbon is the darling of many chemists — organic chemists in particular. The reasons for infatuation with carbon are many. Compounds based on carbon are abundant, diverse, and capable of an amazing array of chemical tricks. Organic chemists like to build structures with carbon building blocks, and then decorate those structures with a variety of other chemical characters. Biology has also found carbon compounds useful, because all life (on Earth, anyway) depends on organic chemistry. In this part, we describe some of the methods and splendors of the chemistry of carbon-based compounds.

Chapter 21

Making Chains with Carbon

In This Chapter

▶ Fueling your knowledge of hydrocarbons

▶ Naming alkanes, alkenes, and alkynes

▶ Saturating and unsaturating hydrocarbons

Any study of organic chemistry begins with the study of *hydrocarbons.* Hydrocarbons are some of the simplest and most important organic compounds. *Organic compounds* are based on carbon skeletons. Hydrocarbon skeletons can be modified — you can dress them up with chemically interesting atoms like oxygen, nitrogen, halogens, phosphorus, silicon, or sulfur. This cast of atomic characters may seem like a rather small subset of the more than 100 elements in the periodic table. It's true: Organic compounds typically use only a very small number of the naturally occurring elements. Yet these molecules include the most biologically important compounds in existence. As an introductory chemistry student, you won't be expected to know more than the basic structure of organic molecules and how to name them. So relax and get organic.

Single File Now: Linking Carbons into Continuous Alkanes

The simplest of the hydrocarbons fall into the category of *alkanes.* Alkanes are chains of carbon molecules connected by single covalent bonds. Chapter 5 describes how single covalent bonds result when atoms share pairs of valence electrons. Carbon molecules have four valence electrons. So, carbon atoms are eager to donate their four valence electrons to covalent bonds so they can receive four donated electrons in turn, filling their valence shell. In other words, carbon really likes to form four bonds. In alkanes, each of these is a single bond with a different partner.

As the name *hydrocarbon* suggests, these partners may be hydrogen or carbon. The simplest of the alkanes, called *continuous-* or *straight-chain alkanes,* consist of one straight chain of carbon atoms linked with single bonds. Hydrogen atoms fill all the remaining bonds. Other types of alkanes include closed circles and branched chains, but we begin with straight-chain alkanes because they make clear the basic strategy for naming hydrocarbons. From the standpoint of naming, the hydrogen atoms in a hydrocarbon are more or less "filler atoms." Alkanes' names are based on the largest number of consecutively bonded carbon atoms. So, the name of a hydrocarbon tells you about that molecule's structure.

TIP

To name a straight-chain alkane, simply match the appropriate chemical prefix with the suffix –*ane*. The prefixes relate to the number of carbons in the continuous chain and are listed in Table 21-1.

Table 21-1		The Carbon Prefixes	
# of Carbons	*Prefix*	*Chemical Formula*	*Alkane*
1	Meth–	CH_4	Methane
2	Eth–	C_2H_6	Ethane
3	Prop–	C_3H_8	Propane
4	But–	C_4H_{10}	Butane
5	Pent–	C_5H_{12}	Pentane
6	Hex–	C_6H_{14}	Hexane
7	Hept–	C_7H_{16}	Heptane
8	Oct–	C_8H_{18}	Octane
9	Non–	C_9H_{20}	Nonane
10	Dec–	$C_{10}H_{22}$	Decane

The naming method in Table 21-1 tells you how many carbons are in the chain. Because you know that each carbon has four bonds and because you are fiendishly clever, you can deduce the number of hydrogen atoms in the molecule as well. Consider the carbon structure of pentane, for example, shown in Figure 21-1.

Figure 21-1:
Pentane's
carbon
skeleton.

Only four carbon-carbon bonds are required to produce the five-carbon chain of pentane. This leaves many bonds open — two for each interior carbon and three for each of the terminal carbons. These open bonds are satisfied by carbon-hydrogen bonds, thereby forming a hydrocarbon, as shown in Figure 21-2.

Figure 21-2:
Pentane's
hydrocarbon
structure.

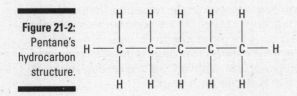

If you add up the hydrogen atoms in Figure 21-2, you get 12. So, pentane contains 5 carbon atoms and 12 hydrogen atoms.

As the organic molecules you study get more and more complicated, it will become more and more important to draw the molecular structure to visualize the molecule. In the case of straight-chain alkanes, the simplest of all organic molecules, you can remember a convenient formula for calculating the number of hydrogen atoms in the alkane without actually drawing the chain:

Number of hydrogen atoms = (2 × number of carbon atoms) + 2

You can refer to the same molecule in a number of different ways. For example, you can refer to pentane by its name (ahem . . . *pentane*), by its molecular formula, C_5H_{12}, or by the complete structure in Figure 21-2. Clearly, these different names include different levels of structural detail. A *condensed structural formula* is another naming method, one that straddles the divide between a molecular formula and a complete structure. For pentane, the condensed structural formula is $CH_3CH_2CH_2CH_2CH_3$. This kind of formula assumes that you understand how straight-chain alkanes are put together. Here's the lowdown:

✔ Carbons on the end of a chain, for example, are only bonded to one other carbon, so they have three additional bonds that are filled by hydrogen and are labeled as CH_3 in a condensed formula.

✔ Interior carbons are bonded to two neighboring carbons and have only two hydrogen bonds, so they're labeled CH_2.

Your chemistry teacher will require you to draw structures of alkanes, given their names, and will require you to name alkanes, given their structure. If your teacher fails to make such requests, ask to see his credentials. You may be dealing with an impostor.

Q. What is the name of the following structure, and what is its molecular formula?

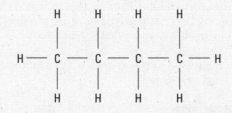

A. **Butane; C_4H_{10}.** First, count the number of carbons in the continuous chain. There are four. Table 21-1 helpfully points out that four-carbon chains earn the prefix *but–*. What's more, this molecule is an alkane (because it contains only single bonds), so it receives the suffix *–ane*. So, what you've got is butane. With four carbon atoms in a straight chain, ten hydrogen atoms are required to satisfy all the carbon bonds, so the molecular formula of butane is C_4H_{10}.

1. What is the name of the following structure, and what is its molecular formula?

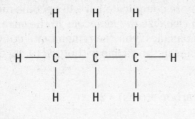

Solve It

2. Draw the structure of straight-chain octane.

Solve It

Going Out on a Limb: Making Branched Alkanes by Substitution

Not all alkanes are straight-chain alkanes. That would be too easy. Many alkanes are so-called *branched alkanes*. Branched alkanes differ from continuous-chain alkanes in that carbon chains substitute for a few hydrogen atoms along the chain. Atoms or other groups (like carbon chains) that substitute for hydrogen in an alkane are called *substituents*.

Naming branched alkanes is slightly more complicated, but you need only to follow a simple set of steps to arrive at a proper (and often lengthy) name.

1. **Count the longest continuous chain of carbons.**

Tricky chemistry teachers often draw branched alkanes with the longest chain snaking through a few branches instead of obviously lined up in a row. Consider the two carbon structures shown in Figure 21-3. The two are actually the same structure, drawn differently! Yikes. In either case, the longest continuous chain in this structure has eight carbons.

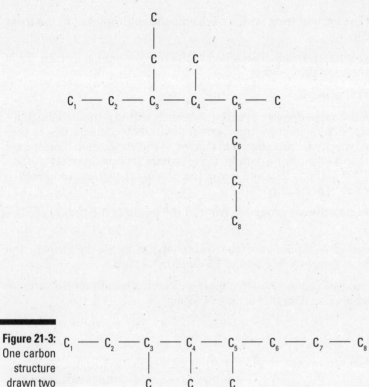

Figure 21-3:
One carbon
structure
drawn two
different
ways.

2. **Number the carbons in the chain *starting with the end that's closest to a branch.***

You can always check to be sure you've done this step correctly by numbering the carbon chain from the opposite end as well. The correct numbering sequence is the one in which the substituent branches extend from the lowest-numbered carbons. For example, as it's drawn and numbered in Figure 21-3, the alkane has substituent groups branching off of its third, fourth, and fifth carbons. If the carbon chain had been numbered backwards, these would be the fourth, fifth, and sixth carbons in the chain. Because the first set of numbers is lower, the chain is numbered properly. The longest chain in a branched alkane is called the *parent chain*.

3. **Count the number of carbons in each branch.**

These groups are called *alkyl groups* and are named by adding the suffix –yl to the appropriate alkane prefix (Table 21-1 awaits your visit). The three most common alkyl groups are the methyl (one carbon), ethyl (two carbons), and propyl (three carbons) groups. Figure 21-3 has two methyl groups, one ethyl group, and no propyl groups.

Be careful when you find yourself dealing with alkyl groups made up of more than just a few carbons. A tricky drawing may cause you to misnumber the parent chain!

4. **Attach the number of the carbon from which each substituent branches to the front of the alkyl group name.**

For example, if a group of two carbons is attached to the third carbon in a chain, like it is in Figure 21-3, the group is called 3-ethyl.

5. **Check for repeated alkyl groups.**

If multiple groups with the same number of carbons branch off the parent chain, don't repeat the name. Rather, include multiple numbers, separated by commas, before the alkyl group name. Also, specify the number of instances of the alkyl group by using the prefixes *di–*, *tri–*, *tetra–*, and so on. For example, if one-carbon groups (in other words, methyl groups) branch off carbons four and five of the parent chain, the two methyl groups appear together as "4,5-dimethyl."

6. **Place the names of the substituent groups in front of the name of the parent chain *in alphabetical order.***

Prefixes like *di–*, *tri–*, and *tetra–* don't figure into the alphabetizing. So, the proper name of the organic molecule in Figure 21-3 is 3-ethyl-4,5-dimethyloctane.

Note that hyphens are used to connect all the naming elements except for the last connection to the parent chain (. . . dimethyl-octane is wrong).

Q. Name the branched alkane shown in the following structure:

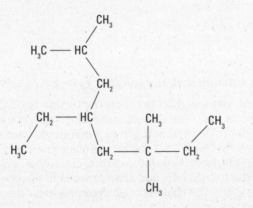

A. **4-ethyl-2,6,6-trimethyloctane.** First, notice how some of the bonds seem to zig-zag. This is a feature of many hydrocarbon structures. The longest continuous chain of carbon atoms in this alkane is eight carbons long. So, the parent chain is octane. Four alkyl groups branch off the parent chain: An ethyl group branches off the fourth carbon, two methyl groups branch off the sixth carbon, and another methyl group branches off the second carbon. Attaching appropriate prefixes and alphabetizing yields the name 4-ethyl-2,6,6-trimethyloctane.

3. Name the branched alkane shown in the following structure:

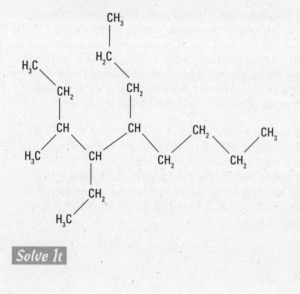

4. Name the branched alkane shown in the following structure:

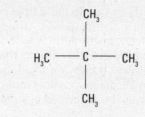

5. Draw the alkane 3,4-diethyl-5-propyldecane.

6. Draw the alkane 3-propyl-2,2,4,4-tetra-methylheptane.

Getting Unsaturated: Alkenes and Alkynes

Carbons can do more than enagage in four single bonds. There's more to organic molecules than substituent-for-hydrogen swaps. When carbons in an organic compound fill their valence shells entirely with single bonds, we say the compound is *saturated*. But many hydrocarbons contain carbons that bond to each other more than once, creating double or triple covalent bonds. We say these hydrocarbons are *unsaturated* because they have fewer than the maximum possible number of hydrogens or substituents. For every additional carbon-carbon bond formed in a molecule, two fewer covalent bonds to hydrogen are formed.

When neighboring carbons share four valence electrons to form a double bond, the resulting hydrocarbon is called an *alkene*. Alkenes are characterized by these chemically interesting double bonds, which are more reactive than single carbon-carbon bonds (see Chapter 5 for a review of sigma and pi bonding). Double bonds also change the shape of a hydrocarbon, because the sp^2 hybridized valence orbitals assume a trigonal planar geometry, as shown by the carbons of ethene in Figure 21-4. Saturated carbon is sp^3 hybridized and has tetrahedral geometry (again, see Chapter 5 to review hybridization).

Figure 21-4:
Ethene, a
two-carbon
alkene.

Naming alkenes is slightly more complicated than naming alkanes. In addition to the number of carbons in the main chain and any branching substituents, you must also note the location of the double bonds in an alkene and incorporate that information into the name. Nevertheless, the essential naming strategy for alkenes is quite similar to that for alkanes in the previous section:

1. **Locate the longest carbon chain, and number it,** *starting at the end closest to the double bond.*

   In other words, double bonds trump substituents when it comes to numbering the parent chain. Build the name of the parent chain by using the same prefixes as used for alkanes (refer to Table 21-1), but match the prefix with the suffix *–ene*. A three-carbon chain with a double bond, for example, is called "propene."

2. **Number and name substituents that branch off the alkene** *in the same way that you do for alkanes.*

   List the number of the substituted carbon, followed by the name of the substituent. Separate the substituent number and name with a hyphen.

3. **Identify the lowest numbered carbon that participates in the double bond, and put that number** *between the substituent names and the parent chain name* **(sandwiched by hyphens), but after all the substituent names.**

   For example, if the second and third carbons of a five-carbon alkene engage in a double bond, then the molecule is called 2-pentene, not 3-pentene. If that same molecule has a methyl substituent at the third carbon, then the molecule is called 3-methyl-2-pentene.

TIP

Alternately, and especially when there are substituents present, the position of an unsaturation is indicated between the prefix and suffix of the parent chain name. So, 3-methyl-2-pentene may also be written 3-methylpent-2-ene.

Alkynes are hydrocarbons in which neighboring carbons share six electrons to engage in triple covalent bonds. The naming strategy for alkynes is the same as that for alkenes, except that the alkyne parent chain is named by matching the prefix with the suffix –*yne*.

TIP

The trick to drawing hydrocarbons is to start at the end of the name and work backwards. The prefix preceding the –*ane*, –*ene*, or –*yne* ending always tells you how many carbons are in the longest chain, so begin by drawing that parent chain. From there, work through the substituent groups, tacking them on as you go. Finally, add hydrogens into the structure wherever there are empty bonds, and voilá! A portrait of a hydrocarbon.

EXAMPLE

Q. Name the following alkene and the following alkyne:

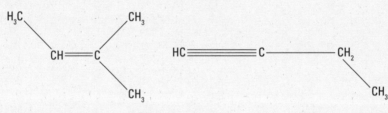

A. **The alkene is 2-methyl-2-butene (or 2-methylbut-2-ene); the alkyne is 1-butyne.** The figure on the left shows a four-carbon alkene with the double bond between the second and third carbon atoms. Numbering the chain from either side yields the same numbers with respect to the double bond. However, numbering from right to left gives a lower number to the methyl substituent, so the compound is 2-methyl-2-butene (or 2-methylbut-2-ene). The figure on the right shows a four-carbon alkyne, with the triple bond located between the first and second carbon atoms. There are no substituents. The compound is therefore 1-butyne.

---

7. Name the following unsaturated hydrocarbon:

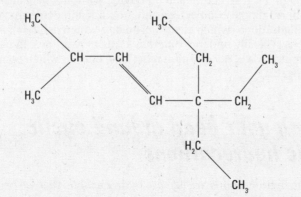

8. Draw the compound 5-ethyl-5-methyl-3-octyne (or 5-ethyl-5-methyloct-3-yne).

9. Draw the compound 4,4-dimethyl-2-pentene (or 4,4-dimethylpent-2-ene).

Solve It

Rounding 'em Up: Circular Carbon Chains

The compounds we cover earlier in this chapter are linear or branched. However, hydrocarbons can be circular, or *cyclical*. Among the cyclical carbons, there are two important categories, the *cyclic aliphatic* hydrocarbons and the *aromatic* hydrocarbons.

Chemists sometimes divide hydrocarbons into aliphatic and aromatic categories to highlight important differences in structure and reactivity. Without going into more technical detail than is useful here, aliphatic molecules and aromatic molecules have significantly different electronic configurations (which electrons go into which orbitals). As a result, the two types of hydrocarbons typically undergo different kinds of reactions. In particular, they tend to undergo different kinds of substitution reactions, ones in which some atom or group substitutes for hydrogen.

Wrapping your head around cyclic aliphatic hydrocarbons

Cyclic aliphatic hydrocarbons are like the hydrocarbons that we explain earlier in this chapter, except that they form a closed ring. The rules for naming these compounds build on the rules we provide earlier in this chapter. For example, a cyclical six-carbon alkane includes the name *hexane,* but is preceded by the prefix *cyclo–,* making the final name *cyclohexane.*

A single substituent or unsaturation on a cyclic aliphatic hydrocarbon doesn't require a number. So, a single bromine-for-hydrogen substitution on cyclohexane yields a compound with the name bromocyclohexane. Likewise, a lone double bond unsaturation on cyclo-hexane yields a compound with the name cyclohexene.

Multiple substitutions or unsaturations require numbering. In these cases, the same rules apply for deciding the rank of substituents. Triple bonds outrank double bonds. Double bonds outrank other substituents. So, number the carbons in the way that respects these rankings and produces the lowest overall numbers. A cyclohexane molecule with two methyl substituents on neighboring carbons, for example, is called 1,2-dimethylcyclohexane.

Sniffing out aromatic hydrocarbons

Aromatic hydrocarbons have special properties because of their electronic structure. Aromatics are *both cyclic and conjugated.* Conjugation results from an alternation of double or triple bonds with single bonds. Noncyclic hydrocarbons can be conjugated, too, but they can't be aromatic. Aromatic molecules have clouds of *delocalized pi electrons,* electrons that move freely through a set of overlapping *p* orbitals. The model aromatic compound is benzene, the resonance structures for which are depicted in Chapter 5. Because of its cyclical, conjugated bonding, pi electrons delocalize evenly into rings above and below the plane of the flat benzene molecule. Aromatic compounds are very stable compared to their aliphatic counterparts.

Numbering substituents on aromatics follows the same basic pattern as followed for cyclic aliphatic compounds. A single substituent requires no numbering, as in bromobenzene. Multiple substituents are numbered by rank, with the highest-ranked substituent placed on carbon number one, and proceeding in a way that results in the lowest overall numbers. A benzene ring with chlorine and methyl substituents situated two carbons away from one another, for example, would be called 1-chloro-3-methylbenzene.

0. Name the following cyclic hydrocarbon:

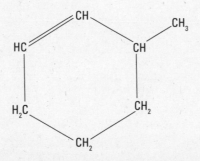

A. **3-methylcyclohexene.** The structure is a six-carbon cyclic alkene with a methyl substituent. Numbering starts with the highest priority group, which is the double bond. Number so that the carbons of the double bond receive numbers 1 and 2, and the carbon to which the methyl group is attached gets the lowest possible number. This means numbering counter-clockwise as shown in the figure. The name of the compound is 3-methylcyclohexene. You don't need to specify the number of the carbon where the double bond appears; because it's the highest-ranking feature, the double bond is assumed to start at carbon number one.

10. Name the following cyclic hydrocarbon:

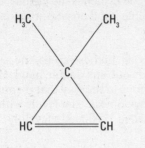

11. Draw the compound 4-methylcyclopentene.

Solve It

12. Draw the compound 4-butyl-3-ethylcyclopentyne.

Solve It

Answers to Questions on Carbon Chains

Do you feel like you're on the straight and narrow path to understanding carbon chains, or do you feel like you've been running around in circles? Take a deep breath, relax, and check your answers to the practice problems presented in this chapter.

1 **The structure is propane; its molecular formula is C_3H_8.** The figure shows a three-carbon chain with only single bonds. Therefore, it is propane. Its molecular formula is C_3H_8.

2 The *oct–* prefix here tells you that this alkane is eight carbons long. Draw eight linked carbons and fill in the empty bonds with hydrogen. Your structure should look like this one:

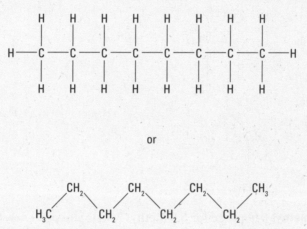

or

3 **4-ethyl-3-methyl-5-propylnonane**. The structure shown is a nine-carbon alkane, so its parent chain name is nonane. Begin numbering from the end that gives the substituent groups the lowest numbers. A one-carbon group extends from the third carbon (3-methyl), a two-carbon group extends from the fourth carbon (4-ethyl), and a three-carbon group extends from the fifth carbon (5-propyl). Alphabetize these substituents and tack on the nonane ending, and you have 4-ethyl-3-methyl-5-propylnonane.

4 **2,2-dimethylpropane.** Note that this is the proper *systematic* name, and that other *common* names may also be used. The common name for this structure is neopentane.

5 The name 3,4-diethyl-5-propyldecane tells you first and foremost that this compound is a ten-carbon alkane. The substituent names indicate two ethyl groups extending from the third and fourth carbons and a propyl group extending from the fifth carbon in the chain. So your art-work should look like this figure:

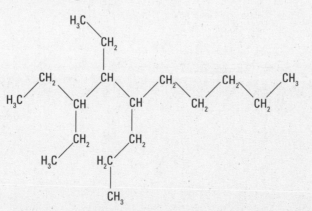

6 The *heptane* ending tells you that you're dealing with a seven-carbon alkane here, so begin by drawing a seven-carbon parent chain. Next, attach the substituents to the main chain by decoding their names. *3-propyl* tells you that a three-carbon substituent group extends from the third carbon in the chain, and *2,2,4,4-tetramethyl* tells you that four single-carbon substituent groups are attached to the chain, two on the second carbon and two on the fourth. Your drawing should look like this:

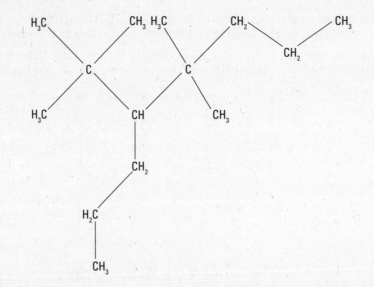

7 **5,5-diethyl-2-methyl-3-heptene (or 5,5-diethyl-2-methylhept-3-ene).** Note first that this structure contains a double bond, meaning it's an alkene and will end with the suffix *–ene*. It has seven carbons in the parent chain, so its base name is *heptene*. In general, you begin numbering the chain at the end closest to the double bond because it has the highest priority. However, in this case, the double bond is between carbons three and four (so it's 3-heptene) no matter which end you begin numbering from. The substituents serve as the tiebreaker here to help you decide whether to number from the left or the right. Choose the end that yields the lowest substituent numbers and then name the substituents. You should get 5,5-diethyl-2-methyl-3-heptene (or 5,5-diethyl-2-methylhept-3-ene).

8 The ending *3-octyne* tells you that this compound is an eight-carbon alkyne with the triple bond between the third and fourth carbons. The substituent names *5-ethyl* and *5-methyl* tell you that the chain includes both a two-carbon (ethyl) and a one-carbon (methyl) branch off the fifth carbon. If your drawing looks like the following figure, you've done well.

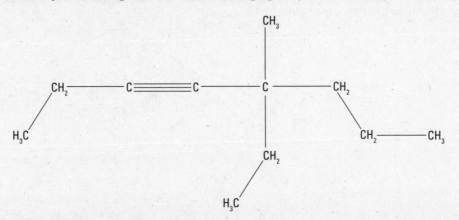

9   The ending *3-pentene* tells you that this compound is a five-carbon alkene with the double bond between the second and third carbons. The substituent name *4,4-dimethyl* tells you that two single-carbon substituent groups are present, both extending from the fourth carbon in the chain. Your artwork should look like the following figure:

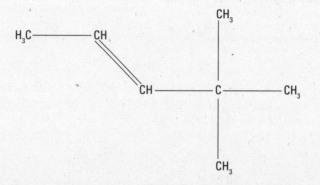

10   **3,3-dimethylcyclopropene.** This is a five-carbon cyclic alkene with two methyl substituents. Number the ring so the two carbons sharing the double bond receive the numbers 1 and 2, and the methyl substituents receive the lowest number. Doing so causes the two methyl groups to extend from the carbon numbered 3. The name is therefore 3,3-dimethylcyclopropene.

11   The ending *cyclopentene* indicates that this is a five-carbon cyclic alkene. Draw a five-carbon ring containing a single double bond and number around the ring so the two carbons sharing the double bond receive the numbers 1 and 2. Then attach a one-carbon substituent group to the carbon numbered 4. Here's what your drawing should look like:

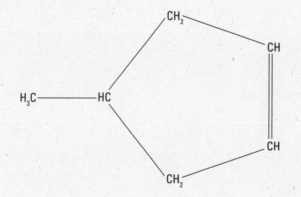

12 The name 4-butyl-3-ethylcyclopentyne reveals that the compound is a five-carbon cyclic alkyne, so begin by drawing a five-carbon ring with one triple bond. Number around the ring so the two carbons sharing the triple bond receive the numbers 1 and 2, and then attach a four-carbon (butyl) substituent to the fourth carbon and a two-carbon (ethyl) substituent to the third carbon. Your drawing should look like this one:

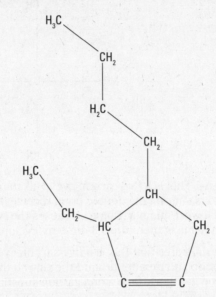

Chapter 22

Seeing Isomers in Stereo

. . . .

In This Chapter

▶ Assigning *cis–* and *trans–* configurations

▶ Reflecting on enantiomers and diastereomers

. . . .

How about this: Two organic molecules have identical chemical formulas. Each atom in one molecule is bonded to the same groups as in the other. They're identical molecules, right? Wrong! (Mischievous chemistry gods point and snicker.) Many organic molecules are *isomers,* compounds with the same formula and types of bonds, but with different structural or spatial arrangements. Who cares about such subtle differences? Well, you might. Consider thalidomide, a small organic molecule widely prescribed to pregnant women in the late 1950s and early 1960s as a treatment for morning sickness. Thalidomide exists in two isomeric forms that rapidly switch from one to the other in the body. One isomer is very effective at combating morning sickness. The other isomer causes serious birth defects. Isomers matter.

Isomers can be confusing. They fall into different categories and subcategories. So, before committing your brain to a game of isomeric Twister, peruse the following breakdown:

✔ *Structural isomers* have identical molecular formulas but differ in the arrangement of bonds.

✔ *Stereoisomers* have identical connectivities — all atoms are bonded to the same types of other atoms — but differ in the arrangement of atoms in space.

    • *Diastereomers* are stereoisomers that are *not* non-superimposable mirror images of each other. Two types of diastereomers exist: *Geometric isomers* (or *cis-trans isomers*) are diastereomers that differ in the arrangement of groups around a double bond, or the plane of a ring. *Conformers* and *rotamers* are diastereomers that differ because of rotations about individual bonds (we don't cover them in this book because they're beyond the scope of general chemistry).

    • *Enantiomers* are stereoisomers that *are* non-superimposable mirror images of each other.

Chapter 21 neatly handles structural isomers, describing how to recognize and name them appropriately. This chapter focuses on the trickier category: stereoisomers.

Picking Sides with Geometric Isomers

Geometric isomers or *cis-trans isomers* are a good place to start in the world of stereoisomers because they're the easiest of the stereoisomers to understand. In the following sections, we explain how isomers relate to alkenes, alkanes that aren't straight-chain, and alkynes (see Chapter 21 for an introduction to these structures).

Alkenes: Keen on cis-trans configurations

Straight-chain alkanes are immune from geometric isomerism because their carbon-carbon single bonds can rotate freely. *Unsaturate* (or add another bond to) one of those bonds, however, and you've got a different story. Alkenes have double bonds that resist rotation. Furthermore, the sp^2 hybridization of double-bonded carbons gives them trigonal planar bonding geometry (see Chapter 5 for an introduction to hybridization). The result is that groups attached to these carbons are locked on one side or the other of the double bond. Convince yourself of this by examining Figure 22-1.

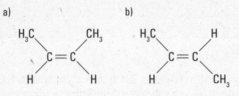

Figure 22-1:
Cis and *trans* isomers of an alkene.

In Figure 22-1a, the carbon chain continues along the same side of the carbon-carbon double bond. Both methyl (-CH₃) groups lie on the same side of the unsaturation. This is called a *cis* configuration. In Figure 22-1b, the carbon chain swaps sides as it proceeds across the double bond. The methyl groups lie on opposite sides of the unsaturation. This is called a *trans* configuration.

Naming *cis-trans* isomers is simple. Attach the appropriate *cis*– or *trans*– prefix before the number referring to the carbon of the double bond (refer to Chapter 21 if you don't know how to name alkenes). For example, Figure 22-1a is *cis*-2-butene, while Figure 22-1b is *trans*-2-butene.

Alkanes that aren't straight-chain: Making a ringside bond

Although straight-chain alkanes happily avoid isomerism by rotating merrily about their single bonds, the four bonds of sp^3-hybridized carbons assume tetrahedral geometry. Detailed representations like the one shown for methane in Figure 22-2 reveal this geometry. In the structure of methane, the bonds depicted as straight lines run in the plane of the page. The bond drawn as a filled wedge projects outward from the page. The bond drawn as a dashed wedge projects behind the page. These filled and dashed wedge symbols are known as *stereo bonds* because they're helpful in identifying stereoisomers.

Figure 22-2:
Stereo bonds in methane.

When alkanes close into rings, they can no longer freely rotate about their single bonds, and the tetrahedral geometry of sp^3-hybridized carbons creates *cis-trans* isomers. Groups bonded to ring carbons are locked above or below the plane of the ring, as shown in Figure 22-3. The figure depicts two different versions of *trans*-1,2-dimethylcyclohexane. In both versions, the adjacent methyl substituents are locked in *trans* positions, on opposite sides of the ring.

> ✔ The upper set of structures shows the plane of the ring as seen from above and highlights the *trans–* configuration of the methyl groups with stereo bonds.

> ✔ The lower set of structures shows the same rings rotated 90 degrees downward and toward you.

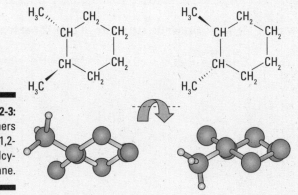

Figure 22-3:
Two isomers of *trans*-1,2-dimethylcy-clohexane.

The *trans* configuration of the methyl groups is most clear in the lower structures. The front-most methyl group is highlighted with explicit hydrogen atoms to emphasize its position above or below the plane of the ring. (Astute readers may ponder, if both are *trans*, why are they different? The answer to this mystery lies in the upcoming "Staring into the Mirror with Enantiomers and Diastereomers" section.)

Alkynes: No place to create stereoisomers

Alkynes also contain carbon-carbon bonds that can't rotate freely. However, the *sp* hybridization of the carbons in these bonds leads to linear bonding geometry. The two-carbon alkyne, ethyne, is shown in Figure 22-4. Each carbon locks three of its valence electrons into the axis of the triple bond. Each has only one valence electron remaining with which to bond to hydrogen. No *cis-trans* isomerism is possible in this scenario.

Figure 22-4:
No isomerism is possible at the triple bonds of alkynes, such as ethyne.

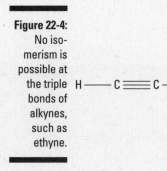

Q. Draw the structure of *cis*-3-hexene.

A. The name helpfully informs you that you're dealing with a six-carbon alkene, and that the double bond occurs between the third and fourth carbon atoms. The *cis*– prefix tells you that the carbon chain continues along the same side of the double bond, as drawn in the following figure:

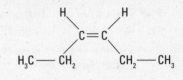

---

1. Name the structure shown in the following figure:

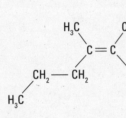

 Solve It

2. Draw the structure of *trans*-2-pentene.

Solve It

---

3. Draw the *cis*– and *trans*– isomers of 3-heptene.

Solve It

4. Can geometric isomerism occur across the first and second carbons in a chain? Draw the structure of 1-butene. Are different geometric isomers possible across the double bond in this molecule? Explain why or why not.

Solve It

Staring into the Mirror with Enantiomers and Diastereomers

Geometric isomers are really just prominent types of *stereoisomers,* compounds that differ only by the arrangement of groups in space. Geometric isomers belong in the category of *diastereomers,* stereoisomers that aren't non-superimposable mirror images. This section introduces a devilish category of stereoisomers called *enantiomers,* isomers that *are* non-superimposable mirror images of each other. Telling the difference between enantiomers and diastereomers can take some practice. You can practice here.

Getting a grip on chirality

Not all mirror images are superimposable on one another. This fact is so fundamental that it may have escaped your attention. If you doubt the truth of it, just try this: Extend your fingers and thumbs so that each hand makes an L-shape. (You've just synthesized two L-shaped molecules — well done.) Now, try to orient your L-shaped hand-molecules so both palms face upwards and so your fingers and thumbs all point in the same directions . . . *at the same time.* It can't be done without serious injury. That's because your L-shaped hand-molecules are *chiral,* meaning they have the property of non-superimposability with their mirror images. Molecules must be chiral in order to be enantiomers.

Carbon atoms can be chiral, too. When sp^3-hybridized carbons bond to four different groups, those carbons have chiral geometry and can form *chiral centers* within molecules. Compare the two molecules shown in Figure 22-5, remembering to visualize the projections of the stereo bonds drawn as wedges (see the earlier section "Alkanes that aren't straight-chain: Making a ringside bond" for more details). Rotate the molecules in your mind, trying to superimpose them. Although this is potentially a safer experiment than trying to superimpose your hands, it stands no greater chance of success. Chirality is what it is.

Figure 22-5:
Each carbon center bonds to the same set of four different partners, but these two chiral carbon atoms are not superimposable.

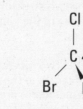

Molecules with chiral centers are often — not always — chiral. Chiral molecules often — not always — possess chiral centers. Chiral centers are important enough that you should understand what they are and seek them out.

Depicting enantiomers and diastereomers in Fischer projections

Chains of tetrahedral carbon atoms show up so frequently in organic molecules that chemists have devised shorthand methods for drawing their structures. One such method is the *Fischer projection.* Fischer projections are a simple way to condense the three-dimensional reality of tetrahedral carbon onto a two-dimensional page.

In a Fischer projection, a bonded chain of tetrahedral carbon atoms is depicted as a vertical line. Horizontal lines represent other bonds projecting from these central carbon atoms. Each intersection between lines represents one carbon atom. Vertical lines symbolize bonds that project away from you. Horizontal lines represent bonds that project toward you. Examples of two molecules drawn as Fischer projections are shown in Figure 22-6.

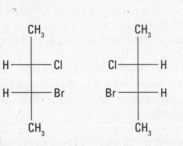

Figure 22-6: Enantiomers of 2-bromo-3-chlorobutane drawn as Fischer projections.

Fischer projections are convenient, but you have to be careful when using them to visualize bonds to make decisions about whether something is chiral.

- ✔ First, you can only rotate these structures *in the plane of the page.* Don't try to rotate them out of the page, or they'll lose all their meaning.

- ✔ Second, *only consider one carbon center at a time.* If you try to simultaneously visualize the three-dimensional bonding of two adjacent carbon atoms on the vertical chain, you'll only get yourself into trouble, and you may well burst a blood vessel.

With those caveats in mind, take a closer look at the two structures in Figure 22-6. Do you see how no amount of sliding or rotating them in the plane of the page can superimpose them? That's because these two molecules are enantiomers.

Compare that situation with the one presented by the two molecules shown in Figure 22-7. Do you see how rotating one of the molecules 180 degrees in the plane of the page allows you to superimpose the two? That's because these two molecules are diastereomers.

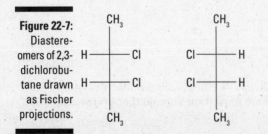

Figure 22-7: Diastereomers of 2,3-dichlorobutane drawn as Fischer projections.

You protest — how can it be? Both sets of molecules contain chiral centers! That's true, but remember: Not all molecules with chiral centers are themselves chiral. In fact, a special term exists for these chiral-but-not-chiral molecules. They are called *meso compounds*. Many meso compounds pull off this trick by having an internal plane of symmetry. In other words, they are their own mirrors. Narcissistic little buggers.

Q. In the structures shown in the following figure, fill in the missing substituents so the two molecules become enantiomers.

molecules are mirror images of each other, as shown in the following figure. The pre-existing substituents guarantee that no pesky *meso* symmetry lurks in the shadows.

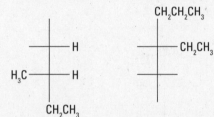

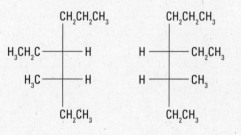

A. To make the two molecules enantiomers, simply fill in the missing pieces so the

---

5. Redraw the structure shown in the following figure as a Fischer projection, identify the chiral centers, and draw the corresponding enantiomer.

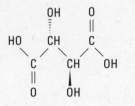

Solve It

6. Which of the structures in the following figure are enantiomers?

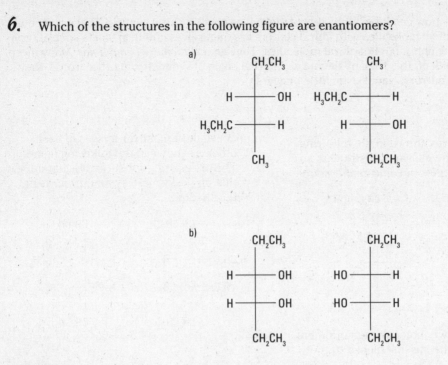

a)

b)

Solve It

7. Which of the structures in the following figure are *meso* compounds?

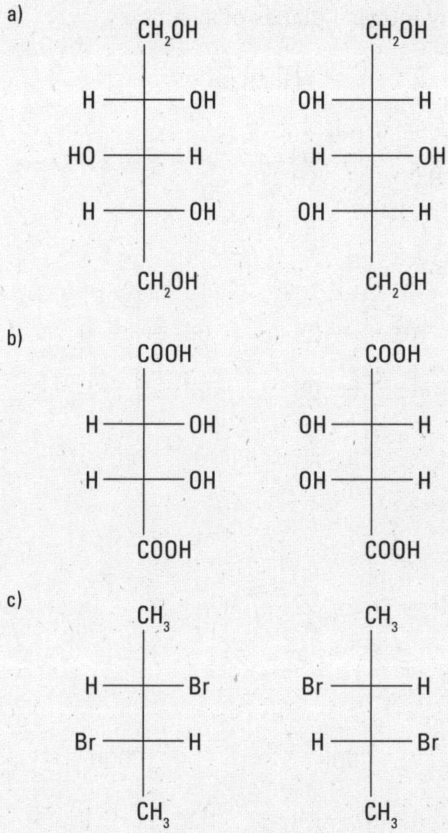

a)

```
      CH₂OH                    CH₂OH
        |                        |
  H ——— OH              OH ——— H
        |                        |
 HO ——— H                H ——— OH
        |                        |
  H ——— OH              OH ——— H
        |                        |
      CH₂OH                    CH₂OH
```

b)

```
      COOH                     COOH
        |                        |
  H ——— OH              OH ——— H
        |                        |
  H ——— OH              OH ——— H
        |                        |
      COOH                     COOH
```

c)

```
      CH₃                      CH₃
        |                        |
  H ——— Br              Br ——— H
        |                        |
 Br ——— H                H ——— Br
        |                        |
      CH₃                      CH₃
```

Solve It

8. Identify the structures in the following figure as geometric isomers, enantiomers, or *meso* compounds. Name any geometric isomers (including *cis–* or *trans–* designations), identify all chiral carbon atoms, and draw any internal planes of symmetry.

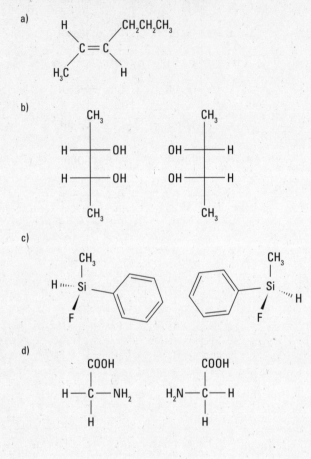

Answers to Questions on Stereoisomers

1 **4-methyl-*trans*-4-nonene.** The structure contains a nine-carbon parent chain, with a double bond occurring between carbon atoms 4 and 5. The main chain swaps sides of the unsaturation as it proceeds. The molecule is also substituted at the fourth carbon with a methyl group.

2 *Trans*-2-pentene is a five-carbon alkene with the double bond occurring between the second and third carbon atoms. The parent chain proceeds along opposite sides of the carbon-carbon double bond, as shown in the following figure:

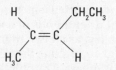

3 The *cis*– and *trans*–isomers of 3-heptene differ only in the orientation of the carbon chain about the double bond, as shown in the following figure:

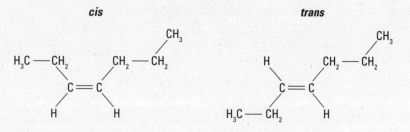

4 Consider the structure of 1-butene, shown in the following figure:

The chain terminates at the double bond. The first carbon atom is bonded to two identical hydrogen atoms. No unique substituent group lies *cis* or *trans* to the continuing parent chain.

5 This is partially a trick question. (Hey! At least we admitted it.) Although the structure has chiral centers, proper drawing of the Fischer projection (pay attention to stereo bonds!), seen in the following figure, makes it clear that the compound is *meso*. The molecule has an internal plane of symmetry.

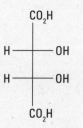

6   **The two structures shown in Figure A are enantiomers,** although they don't appear to be at first glance. To make the enantiomeric relationship more obvious, rotate one of them 180 degrees in the plane of the page. **The two structures in Figure B are not enantiomers.** In fact, this is a *meso* compound.

7   **The structures shown in Figures A and B are *meso* compounds. The structures shown in Figure C are not *meso*.** Although the structures in Figure A don't have an internal plane of symmetry, the two are superimposable mirror images. The structures in Figure B have an internal plane of symmetry and, in fact, are the same compound. The two structures in Figure C are non-superimposable mirror images, and are therefore enantiomers.

8   **The structure in Figure A is a geometric isomer, *trans*-2-hexene. The pair of molecules shown in Figure B are different orientations of a *meso* compound with two chiral centers and an internal plane of symmetry, as shown in the following figure.** The plane cuts the molecule in half, across the bond connecting the second and third carbon atoms. **The pair of structures in Figure C are enantiomers.** The chiral center in this case isn't carbon! Rather, it's silicon, which sits right below carbon on the periodic table, and also bonds with tetrahedral geometry. **The pair of molecules shown in Figure D aren't isomers at all,** but are simply different orientations of the amino acid glycine. The central carbon of the projection bonds to two identical hydrogen atoms, so no isomerism is possible.

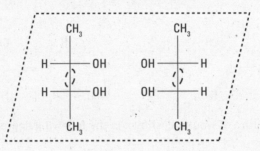

Chapter 23

Moving through the Functional Groups

. . . .

In This Chapter

▶ Beyond hydrocarbons: Recognizing the key organic functional groups

▶ Connecting monomers to make polymers

▶ The wide world of biochemistry: Understanding carbohydrates, proteins, and nucleic acids

. . . .

*I*t's no coincidence that in this, the final chapter that covers new material, we introduce two of the most fascinating and relevant branches of modern chemistry: organic chemistry and biochemistry. We have chosen to do this not because we're malevolent human beings who want to rub in that you still have an awfully long way to go in your study of chemistry, but rather because we want to point out that the skills and information that you have gained from this book open the door to a whole new world of science.

What do organic chemistry and biochemistry have in common? Much of biochemistry *is* organic chemistry, making and breaking bonds in carbon-based compounds. Many compounds involved in biochemistry are based on hydrocarbon skeletons, but are embellished with *functional groups*. These groups include noncarbon elements and give organic compounds a dizzying array of different properties. In this chapter, we explain the basics of how functional groups operate in organic chemistry and biochemistry.

You already know a little about organic chemistry from Chapters 21 and 22, and surely you've noticed that those chapters cover compounds made entirely of hydrogen and carbon. Rest assured that we go beyond hydrocarbons in this chapter! Because hydrocarbons are old news, we sometimes refer to them in this chapter simply by using the letter R, representing any hydrocarbon group: branched or unbranched, saturated or unsaturated.

Meeting the Cast of Chemical Players

In Chapters 21 and 22, you focus intently on hydrocarbons. In fact, you process so much information about those amazing compounds in the previous two chapters that you may forget that the periodic table contains more than 100 other elements. Although organic chemistry often concentrates on just a few of those elements, you need to be familiar with many other compound types, and this time they're made of more than just hydrogen and carbon. All of these exotic new compounds contain plain vanilla carbon chains, but each has a distinguishing feature composed of another, more exotic element.

Table 23-1 summarizes the compounds in the following sections, their distinguishing features, and their endings. Some molecules, including some of the biological monomers that are introduced in the later section "Seeing Chemistry at Work in Biology," contain more than one of these distinguishing features. Locations where one of these groups or a multiple bond appears are prime sites for reactions to occur. These sites are therefore called *functional groups*. Carefully study the functional group column in Table 23-1 so you can recognize them quickly. These are the structures you should keep an eye out for later in this chapter.

Table 23-1		Functional Groups	
Compound Name	*Compound Formula*	*Functional Group*	*Prefix or Suffix*
Alcohol	R-OH		–ol
Ether	R-O-R		ether
Carboxylic acid	R-COOH		–oic acid
Ester	R-COOR		–oate
Aldehyde	R-CHO		–al
Ketone	R-COR		–one

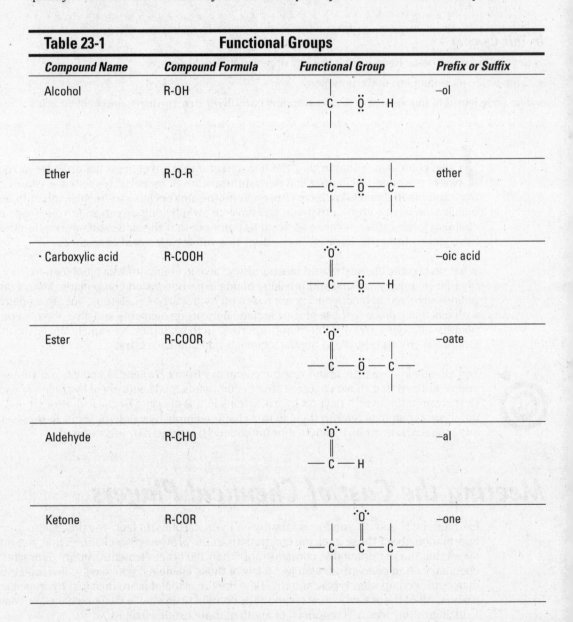

Compound Name	Compound Formula	Functional Group	Prefix or Suffix
Halocarbon	R-X	$-\overset{\displaystyle |}{\underset{\displaystyle |}{C}} - \ddot{\underset{\displaystyle \cdot\cdot}{X}} :$ (X = halogen)	fluoro–, chloro–, bromo–, or iodo–
Amine	R-NH$_2$	$-\overset{\displaystyle |}{\underset{\displaystyle |}{C}} - \overset{\displaystyle \cdot\cdot}{\underset{\displaystyle |}{N}} -$	–amine

Alcohols: Hosting a hydroxide

Alcohols are hydrocarbons with a hydroxyl group attached to them. Their general form is R-OH. To name an alcohol, you simply count the longest number of consecutive carbon atoms in the chain, find its prefix in the table of hydrocarbon prefixes in Chapter 21, and then attach the ending *–ol* to that prefix. For example, a one-carbon chain with an OH group attached is methanol, two is ethanol, three is propanol, and so on.

Substituents (groups branching off the main chain) must, as always, be accounted for, and you must specify the number of the carbon atom in the chain to which the hydroxyl (OH) group is attached. Begin your numbering at the end of the chain closest to the OH, and attach the number before the prefix + *–ol*. For example, the compound shown in Figure 23-1 contains a six-carbon chain with a hydroxyl group. Its base name is therefore hexanol. It has methyl groups (with one carbon each) on the second and fourth carbons and its OH group lies on the third carbon. Its proper name is therefore 2,4-dimethyl-3-hexanol.

Figure 23-1:
An example
of an
alcohol.

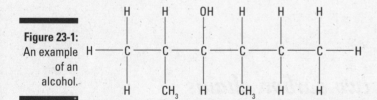

Ethers: Invaded by oxygen

Perhaps most commonly known for their use as early anesthetics, *ethers* are carbon chains that have been infiltrated by an oxygen atom. These oxygen atoms lie conspicuously in the middle of a carbon chain like badly disguised spies in an enemy camp. They have the general formula R-O-R. Ethers are named by naming the alkyl groups on either side of the lone oxygen as substituent groups individually (adding the ending –*yl* to their prefixes), and then attaching the word ether to the end. For example, the compound shown in Figure 23-2 is an ether with a methyl group (one carbon) on one side of the oxygen and an ethyl (two carbons) on the other. Placing the substituents in alphabetical order, the compound's proper name is ethyl methyl ether.

Figure 23-2:
An example
of an ether.

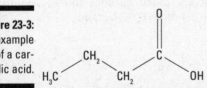

Oxygen atoms in ethers are often surrounded by two identical alkyl groups, in which case the prefix *di*– must be attached to the name of that substituent. For example, an oxygen surrounded by two propyl groups (with three carbons each) is called dipropyl ether.

Carboxylic acids: –COOH brings up the rear

Carboxylic acids appear to be ordinary hydrocarbons until you reach the very last carbon in their chain, whose three ordinary hydrogens have been usurped by a double-bonded oxygen and a hydroxyl group (an OH group). This gives a carboxylic acid the general form R-COOH. These compounds are named by attaching the suffix –*oic acid* to the end of the prefix. For example, the compound shown in Figure 23-3 has four carbons, the last in the chain attached to a double-bonded oxygen and a hydroxyl group. The compound is therefore called butanoic acid. The hydrogen of the COOH group can pop off as H^+, which is why these compounds are called acids.

Figure 23-3:
An example
of a car-
boxylic acid.

Esters: Creating two carbon chains

What if the same double-bonded oxygen/hydroxyl pair from a carboxylic acid has infiltrated deeper into the hydrocarbon chain and is not on an end, but rather in the middle? In order to accomplish this, the COOH group must lose the hydrogen of its hydroxyl group (which, as an acidic hydrogen, is no problem) to open up a bond to which a second hydrocarbon can attach. A compound of this nature, with the general formula R-COOR, is called an *ester*.

Because the carbon chain in an ester is also broken by an oxygen, you need to choose one carbon chain as a lowly substituent group and the other to bear the proud suffix of –*oate*. This high-priority group is always the carbon chain that includes the carbon double bonded to one oxygen and single bonded to the other. The group on the other side of the single-bonded oxygen is named as an ordinary substituent. For example, the compound shown in Figure 23-4 is phenyl ethanoate, because the high-priority group has two carbons and the low-priority group is a benzene ring.

Figure 23-4:
An example
of an ester.

That's right! The low-priority group on the other side of the oxygen doesn't have to be a hydrocarbon chain! It can be a ring or a metal as well. In Figure 23-4, this group is a benzene ring.

Aldehydes: Holding tight to one oxygen

Aldehydes are much like carboxylic acids except that they lack the second oxygen in the COOH group. Their final carbon shares a single bond with its neighboring carbon, a double bond with an oxygen, and a single bond with hydrogen. Aldehydes are named with the suffix *–al* (not to be confused with the *–ol* of alcohols) and have the general formula R-CHO. For example, the compound in Figure 23-5 is pentanal, a five-carbon chain with a double-bonded oxygen taking the place of two of the hydrogen bonds on the final carbon.

Figure 23-5:
An example
of an
aldehyde.

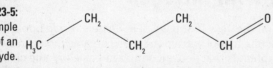

Ketones: Lone oxygen sneaks up the chain

Much like esters are to carboxylic acids, *ketones* share the same basic structure as aldehydes, except that their double-bonded oxygen can be found hiding out in the midst of the carbon chain, giving them the general form R-COR. Ketones are named by adding the suffix *–one* to the prefix generated by counting up the longest carbon chain. Unlike esters, however, the carbon chain of a ketone isn't broken by the double-bonded oxygen group, which makes naming them much simpler. The compound shown in Figure 23-6, for example, is called simply 2-butanone; the number before the name specifies the number of the carbon to which the oxygen is attached.

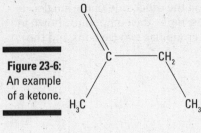

Figure 23-6:
An example
of a ketone.

Halocarbons: Hello, halogens!

Halocarbons are simply hydrocarbons with one or more halogens tacked on in place of a
hydrogen atom. (A *halogen* is a group VIIA element.) Halogens in a halocarbon are always
named as substituent groups — *fluoro–*, *chloro–*, *bromo–*, or *iodo–*, one for each of the four
halogens that form halocarbons (F, Cl, Br, or I). The shorthand for a halocarbon is R-X, where
X is the halogen. The compound shown in Figure 23-7 has two bromo groups attached, one
stemming from the second carbon and one from the third in a five-carbon alkane. Its official
name is, therefore, 2,3-dibromopentane. (See Chapter 21 for more details on naming alkanes.)

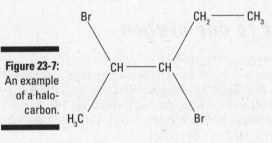

Figure 23-7:
An example
of a halo-
carbon.

Amines: Hobnobbing with nitrogen

Amines are as conspicuous as Waldo on the very first page of a *Where's Waldo* book. No trick-
sters wearing red-striped shirts are waiting to fool you among these organic molecules. Amines
and their derivatives are the only compounds that you encounter in basic organic chemistry
that contain nitrogen atoms. Amines have the basic form R-NH$_2$, and they're named by naming
the carbon chain as a substituent (with the ending *–yl*), and then adding the suffix *–amine*. The
structure in Figure 23-8, for example, is called ethylamine because it's a two-carbon ethyl
chain with an amine group.

Figure 23-8:
An example
of an amine.

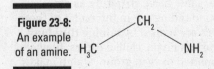

EXAMPLE

Q. Draw the structure and write the condensed formula for 3,4-dimethyl heptanoic acid.

A. **The condensed formula is $CH_3CH_2CH_2CH_3CHCH_3CHCH_2COOH$.** The ending –*oic acid* tells you that you're dealing with a carboxylic acid, and the prefix *hept*– tells you that the carbon chain on the end of the carboxylic acid functional group is seven carbons long. Finally, the 3,4-dimethyl tells you that the molecule has two single-carbon substituent groups, one on the third carbon in the chain and one on the fourth. Remember to number the chain starting from the carbon that is bonded to the oxygens. Add all of these bits of information together and you get the structure shown in the following figure, which has the condensed formula $CH_3CH_2CH_2CH_3CHCH_3CHCH_2COOH$.

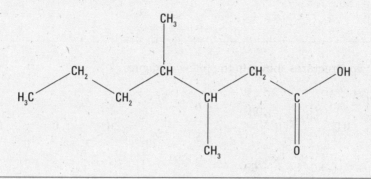

1. Name the two structures shown in the following figure.

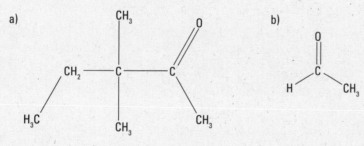

Solve It

2. Draw 4,4,5-triethyl 3-heptanol and 3,4,5-triethyl octanal.

3. Name the two structures shown in the following figure.

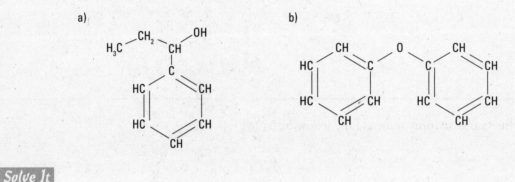

4. Draw ethanoic acid and butyl ethyl ether.

5. Name the structure shown in the following figure.

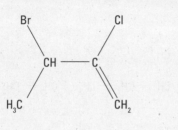

6. Draw 2-methyl butylamine.

Reacting by Substitution and Addition

The carbon-based organic molecules in the previous section can morph into one another through chemical reactions. Such reactions are rather common. Alcohols, for example, are prone to a chemical process called *substitution* in which their OH group is replaced by a different atom. For example, the OH group on 2-pentanol can be replaced by the halogen fluorine, turning the alcohol into a halocarbon called 2-fluoro-pentane (see Figure 23-9).

$$CH_3 - CH_2 - CH_2 - \overset{\overset{\displaystyle OH}{|}}{CH} - CH_3 + F_2 \longrightarrow$$

Figure 23-9: The process of substitution.

$$CH_3 - CH_2 - CH_2 - \overset{\overset{\displaystyle F}{|}}{CH} - CH_3$$

The double bonds of alkenes also make them particularly likely to react with other compounds through a process called *addition*. If the double bond between two carbon molecules is broken, it allows each of those two carbon atoms to form a bond with another atom or molecule. The reaction shown in Figure 23-10, for example, shows water being added across the double bond of 1-hexene. The water molecule itself is split into two pieces — simple hydrogen and a hydroxide — and each of them is added to one of the two carbons that used to share a double bond, forming either 1-hexanol or 2-hexanol. (If you're curious, the product in Figure 23-10 is 2-hexanol.)

Figure 23-10:
The process
of addition.

$$H_2C = CH - CH_2 - CH_2 - CH_2 - CH_3 \longrightarrow H_3C - \overset{\overset{\displaystyle OH}{\mid}}{CH} - CH_2 - CH_2 - CH_2 - CH_3$$

Q. Fill in the blank with the missing reactant in the reaction below.

_____ + $Br_2 \rightarrow CH_3CHBrCH_2CH_3$

A. **The missing reactant is $CH_3CHOHCH_2CH_2CH_3$ (or 2-pentanol).**
A sketch of the structure of the product in this reaction can help you identify the missing reactant.

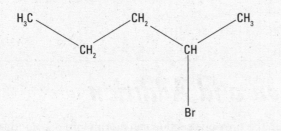

Examine the preceding figure and you see that the substitution occurred on the second carbon in the chain. This is because the bromine group of the product is attached to the second carbon in its chain. One atom or functional group takes the place of another in a substitution reaction, and halogens often take on the role of the replacement atom. A likely reactant, therefore, is one in which the bromo group on the second carbon in the product is replaced by a hydroxyl group. The following figure shows this reactant, 2-pentanol.

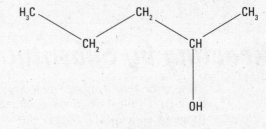

7. Predict the reactants that combined to make the product 3-chlorooctane.

Solve It

8. Predict the product of a reaction combining 1-pentene with water.

Solve It

Seeing Chemistry at Work in Biology

All the compounds that we discuss earlier in this chapter are relatively small molecules. We haven't even given you a prefix for a carbon chain with more than ten carbon atoms. Yet many fundamental organic compounds contain chains of carbons tens and hundreds and thousands long, with lots of hydrogens, nitrogens, and oxygens tacked on along the way. These giant molecules, also known as *polymers,* are built by taking smaller units called *monomers* and joining them together like joining train cars to make a longer train. Just as trains can be constructed with a variety of railcar sequences, polymers can be constructed with a series of identical monomers or with a mixture of two or more different monomers.

Each time another monomer unit is added to a polymer chain, it changes the compound's physical properties, such as stiffness and boiling point. Although they may seem fairly complicated and difficult to make, polymers aren't rare. Your milk jug, shampoo bottle, rubber duckie, and flame-retardant suit are all made of polymers. In fact, all of the compounds that we commonly refer to as plastics are polymers, and manufacturers routinely use chemistry to achieve desirable properties in the plastics they use. Reactions which link monomers to make a polymer are called *polymerization reactions.*

Plastics and rubber aren't the end-all, be-all of polymers however. The human body contains a whole slew of naturally occurring polymers, which you encounter when you look at the nutrition facts on the back of your box of morning cereal. Carbohydrates, proteins, and many lipids are polymers. So too are the DNA and RNA that store and transmit your genetic code. In the following sections, we focus on several important biological polymers: carbohydrates, proteins, and nucleic acids.

Note that in all the reactions in the following sections, the two monomers being linked together lose an oxygen and two hydrogens to create the polymer. Because this essentially amounts to the loss of a water molecule between the two monomers, the process is called *dehydration* or *condensation reaction.*

Carbohydrates: Carbon meets water

Carbohydrates are polymers made up of aldehyde and ketone monomers (see the earlier sections on these two types of compounds). As such, they may contain only the elements carbon, hydrogen, and oxygen. Categories of carbohydrates include starches and sugars, and sugars themselves fall into three categories: *monosaccharides, disaccharides,* and *polysaccharides.* Because *saccharide* is simply a fancy name for sugar, it follows that these compounds are made of one, two, and more than two sugar monomers, respectively. The general formula for a carbohydrate is $C(H_2O)_n$, whether you are dealing with a monosaccharide or a polysaccharide that is hundreds of sugars long.

Glucose, the product of photosynthesis in plants (see Figure 23-11), is arguably the most important biological monosaccharide. Glucose monomers can be combined to make a variety of polysaccharides, including starches (which provide the primary source of carbohydrates in pasta and potatoes) and cellulose (which makes up cell walls in plant cells). Figure 23-12 shows two monosaccharides combining to form a disaccharide.

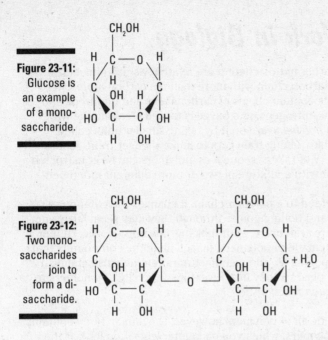

Figure 23-11:
Glucose is an example of a mono-saccharide.

Figure 23-12:
Two mono-saccharides join to form a di-saccharide.

Proteins: Built from amino acids

The monomers that make up proteins are called *amino acids* (see one in Figure 23-13). The amino acid's identity is determined by the structure of the side chain R, which can be any of a variety of organic structures. Individual amino acids are linked together into polymers through the formation of *peptide bonds*. A polymer made of two amino acids is therefore called a *dipeptide* (see Figure 23-14), while more than two amino acids form a *polypeptide*.

Figure 23-13:
An amino acid is the monomer that builds proteins.

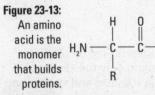

Figure 23-14:
Two amino
acids con-
nected by
a peptide
bond
make up a
dipeptide.

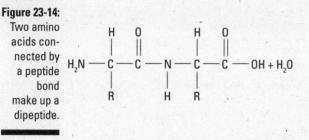

Figure 23-14:
Two amino
acids con-
nected by
a peptide
bond
make up a
dipeptide.

The proteins in the human body (or those in an octopus or a bacterium) are made up of 20 common amino acids. Every protein type is constructed of a unique sequence of amino acids, and your DNA tells your body how to manufacture the proteins it needs by specifying this sequence.

Nucleic acids: The backbones of life

DNA and RNA are extremely important biological polymers. They are constructed of smaller subunits called *nucleotides* (see Figure 23-15), which make up the polymers called *nucleic acids* (see Figure 23-16). Each nucleotide consists of three parts: a phosphoric acid, a sugar, and a nitrogen-containing base. DNA nucleotides use the sugar deoxyribose and employ four types of base. RNA nucleotides use the sugar ribose and employ a nearly identical set of bases. Bases are broken up into two categories:

- *Pyrimidines* are characterized by a single ring of carbon and nitrogen atoms.

- *Purines* contain a double ring.

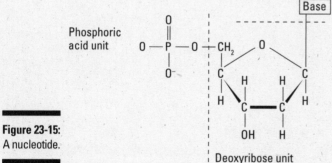

Figure 23-15:
A nucleotide.

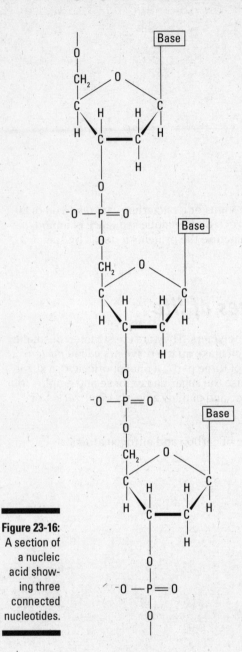

Figure 23-16:
A section of
a nucleic
acid show-
ing three
connected
nucleotides.

The human genome is made up of a sequence of 3 billion nucleotides that tell bodies how to do everything from manufacturing proteins to sneezing.

Q. What types of functional groups are attached to the central ring of the nitrogen-containing base cytosine, shown in the following figure?

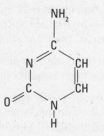

A. The oxygen double bonded to a carbon in the ring indicates a ketone, while the NH$_2$ group attached to a separate carbon in the ring indicates an amine.

9. Is cytosine a purine or a pyrimidine? How can you tell?

Solve It

10. What types of functional groups are present in an amino acid? A carbohydrate?

Solve It

Answers to Questions on Functional Groups

The following are the answers to the practice problems in this chapter.

1   **a. 3,3-dimethyl-2-pentanone.** The longest carbon chain in this structure is five carbons long, and the lone double-bonded oxygen in the middle of the carbon chain tells you that this compound is a ketone. Its basic name is 2-pentanone. It also has two methyl substituents, both off the number three carbon. The proper name for the structure is, therefore, 3,3-dimethyl-2-pentanone.

   b. Ethanoic acid. The second structure also has a double-bonded oxygen; however, it appears on the end of a two-carbon chain. Compound b is the carboxylic acid called ethanoic acid.

2   The name 4,4,5-triethyl 3-heptanol tells you that you're dealing with a seven-carbon alcohol whose hydroxyl group stems from the third carbon in the chain, and that it has three ethyl substituent groups (two off the fourth carbon and one off the fifth). This should lead you to structure a in the following figure. The name 3,4,5-triethyl octanal, on the other hand, is an eight-carbon aldehyde, also with three ethyl substituent groups (this time with one off the third carbon, one off the fourth, and one off the fifth). Check it out in structure b in the following figure.

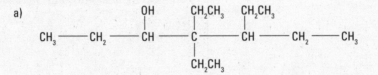

3   **a. 1-phenyl-1- propanol.** The longest carbon chain in this structure is three carbons long. The final carbon in that chain has both an alcohol and a phenyl group extending from it. The alcohol group gives the compound the basic name propanol, and adding the phenyl substituent, you should arrive at the name 1-phenyl-1-propanol.

   b. Diphenyl ether. Structure b is an ether with two identical phenyl groups surrounding the oxygen, making it diphenyl ether.

4   Ethanoic acid is a simple two-carbon carboxylic acid, as shown in structure a of the following figure. Butyl ethyl ether is an ether with a butyl group on one side of the oxygen and an ethyl group on the other, as shown in structure b of the following figure.

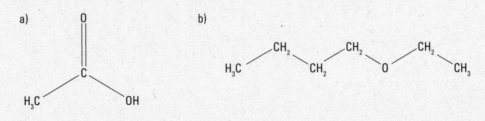

5 **3-bromo-2-chloro-1-butene.** This structure is a halocarbon. The longest carbon chain is four carbons long, and it's an alkene, making the base name butene (see Chapter 21 for details on naming alkenes). Numbering from the side closest to the double bond yields the lowest overall numbers; therefore, the compound is 3-bromo-2-chloro-1-butene.

6 2-methyl butylamine is a four-carbon amine with a methyl substituent off its second carbon. It has the structure shown in the following figure.

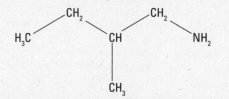

7 **3-octanol and diatomic chlorine.** A halocarbon with a halogen extending from the third carbon of an eight-carbon chain (otherwise known as 3-chlorooctane) was most likely created in a substitution reaction during which 3-octanol reacted with diatomic chlorine to produce 3-chlorooctane. This is true because a substituent hydroxyl group is quite likely to be replaced in a substitution reaction.

8 When 1-pentene is combined with water, an addition reaction is likely to occur in which the double bond of pentene is broken, and a hydrogen and a hydroxide are added to the newly freed carbons, forming either 1-pentanol or 2-pentanol.

9 **Cytosine is a pyrimidine.** You can tell because it contains a single ring of carbon and nitrogen atoms, instead of a double ring.

10 **The amino acid has an amine group and a carboxylic acid group, while the carbohydrate has several alcohol groups and an ether group.** Examining the structure of an amino acid monomer in Figure 23-13 reveals that it contains both an amine group ($-NH_2$) and a carboxylic acid group ($-COOH$). As for carbohydrates, the monosaccharide glucose shown in Figure 23-11 has several alcohol ($-OH$) functional groups and an oxygen infiltrating the carbon ring (an ether group).

Part VI
The Part of Tens

The 5th Wave   By Rich Tennant

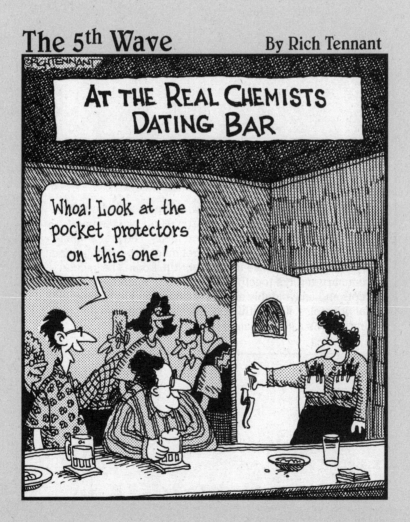

In this part . . .

This part is a lean, hard-edged, white-knuckled distillation of the rest of the book. You'll find two ten-point summaries of key information and ideas. The first summary is particularly useful with its list of many key formulas and equations that you use throughout the book. The second summary gathers together several head-scratching concepts and unavoidably annoying exceptions to basic chemical principles. If something doesn't seem to make sense at first, there's a good chance you'll find it explained here.

Chapter 24

Ten Formulas to Tattoo on Your Brain

or some, the mere presence of an equation is scary. Take consolation in this thought: Understanding an equation saves you brain space because equations pack a whole lot of explanation into one compact little sentence.

For example, say you need to remember what decreasing temperature does to the volume of a gas when pressure is held constant. Maybe you'll just remember. If not, you could try to reason it out based on kinetic theory. That approach may not work for you either. But if you simply remember the Combined Gas Law, you can figure it out simply by inspecting the equation. Not only that, but all other combinations of increasing or decreasing temperature, pressure, and volume fall instantly within your grasp. Equations can be your friends. Your friends have gathered to meet you in this chapter. Hang out for a while.

The Combined Gas Law

Using the following equation for the Combined Gas Law, which we introduce in Chapter 11, you can calculate how a gas responds when one factor changes. Changing any one parameter (temperature, T, pressure, P, or volume, V) affects another parameter when the third parameter is held constant. To work the equation, first cancel out the constant parameter. Then plug in the values for your known parameters. Finally, solve for the unknown.

$$\frac{P_1 \times V_1}{T_1} = \frac{P_2 \times V_2}{T_2}$$

Values with a subscripted 1 refer to initial states, and values with a subscripted 2 refer to final states.

Dalton's Law of Partial Pressures

The following law, which we introduce in Chapter 11, is useful with mixtures of gases at constant volume and temperature. Dalton's Law of Partial Pressures states that the total pressure of the mixture is simply the sum of the partial pressures of each of the individual gases. You can solve for the total pressure (P_{total}) or any individual partial pressure (P_1, P_2, and so on) as long as you know all the other pressures.

$$P_{total} = P_1 + P_2 + P_3 \ldots + P_n$$

The Dilution Equation

The following relation works because the number of moles of a solute doesn't change when you dilute the solution. You can use the Dilution Equation to calculate an initial concentration or volume (M_1, V_1) or a final concentration or volume (M_2, V_2), as long as you know the other three values. See Chapter 12 for more information on dilutions.

$$M_1 \times V_1 = M_2 \times V_2$$

Rate Laws

Rate laws, which we cover in Chapter 14, relate reaction rates to the concentrations of reactants. Which rate law is appropriate depends on the kind of reaction involved. Zero-order reactions have rates that don't depend on reactant concentrations; first-order reactions have rates that depend on the concentration of one reactant; and second-order reactions have rates that depend on the concentrations of two reactants. Here are representative reaction equations for zero-order, first-order, and second-order reactions.

Zero-order:	$A \rightarrow B$	(rate is independent of [A])
First-order:	$A \rightarrow B$	(rate depends on [A])
Second-order:	$A + A \rightarrow C$	(rate depends on [A]), *or*
	$A + B \rightarrow C$	(rate depends on [A] and [B])

The rate laws for these reactions describe the rate at which reactant A disappears over time, $-d[A] / dt$. Note that dX / dt is simply the rate at which X changes at any given moment in time, t. Rate laws are expressed in terms of k, the rate constant:

Zero-order rate law:	rate = $-d[A] / dt = k$
First-order rate law:	rate = $-d[A] / dt = k[A]$
Second-order rate law:	rate = $-d[A] / dt = k[A]^2$ *or*
	rate = $-d[A] / dt = k[A][B]$

The Equilibrium Constant

You can calculate the equilibrium constant for a reaction, K_{eq}, from the concentrations of reactants and products at equilibrium. In the following reaction, for example, A and B are reactants, C and D are products, and *a, b, c,* and *d* are *stoichiometric coefficients* (numbers showing mole multiples in a balanced equation):

aA + bB ↔ cC + dD

The equilibrium constant for the reaction is calculated as:

$$K_{eq} = \frac{\left[C\right]^c \left[D\right]^d}{\left[A\right]^a \left[B\right]^b}$$

Note that [X] refers to the molar concentration of X; $[0.7] = 0.7 \text{ mol L}^{-1}$.

Spontaneous reactions have K_{eq} values greater than 1. Nonspontaneous reactions have K_{eq} values between 0 and 1. The inverse of a K_{eq} value is the K_{eq} for the reverse reaction.

Check out Chapter 14 for more details about the equilibrium constant.

Free Energy Change

The free energy, G, is the amount of energy in a system available to do work. Usually, the reactants and products of a chemical reaction possess different amounts of free energy. So, the reaction proceeds with a change in energy between reactant and product states, $\Delta G = G_{product} - G_{reactant}$. The change in free energy arises from an interplay between the change in enthalpy, ΔH, and the change in entropy, ΔS:

$\Delta G = \Delta H - T\Delta S$ (where T is temperature)

Spontaneous reactions release energy, and occur with a negative ΔG. Nonspontaneous reactions require an input of energy to proceed and occur with a positive ΔG.

The change in free energy associated with a reaction is related to the equilibrium constant for that reaction (see the previous section for more about the equilibrium constant), so you can interconvert between ΔG and K_{eq}:

$\Delta G = -RT\ln K_{eq}$ (R is the gas constant, T is temperature)

Note that "ln" stands for natural log (that is, log base e). You'll find a button for ln on any respectable scientific calculator.

Flip to Chapter 14 for more information about free energy.

Constant-Pressure Calorimetry

Calorimetry is the measurement of heat changes that accompany a reaction (see Chapter 15 for details). The important values to know are heat (q), mass (m), specific heat capacity (C_p), and the change in temperature (ΔT). If you know any three of these values, you can calculate the fourth:

$$q = mC_p\Delta T$$

Be sure that your units of heat, mass, and temperature match those used in your specific heat capacity before attempting any calculations.

Hess's Law

The heat taken up or released by a chemical process is the same whether the process occurs in one or several steps. So, for a multistep reaction (at constant pressure), such as

$$A \rightarrow B \rightarrow C \rightarrow D$$

the heats of the individual steps add up to equal the total heat for the reaction:

$$\Delta H_{A\rightarrow D} = \Delta H_{A\rightarrow B} + \Delta H_{B\rightarrow C} + \Delta H_{C\rightarrow D}$$

So, you can calculate the overall change in heat, or the change in heat for any given step, as long as you know all the other values. This formula is known as Hess's Law.

Moreover, the reverse of any reaction occurs with the opposite change in heat:

$$\Delta H_{A\rightarrow D} = -\Delta H_{D\rightarrow A}$$

This means you can use known heat changes for reverse reactions simply by changing their signs.

Flip to Chapter 15 for the full scoop on Hess's Law.

pH, pOH, and K_W

As we explain in Chapter 16, pH and pOH are measurements of the acidity or basicity of aqueous solutions:

$$pH = -\log[H^+]$$
$$pOH = -\log[OH^-]$$

Pure water spontaneously self-ionizes to a small degree

$$H_2O(l) \leftrightarrow H^+(aq) + OH^-(aq) \qquad \text{(note: } H^+ \text{ exists in water as } H_3O^+)$$

leading to equal concentrations of hydrogen and hydroxide ions (10^{-7} mol L^{-1}). The product of these two concentrations is the ion-product constant for water, K_w:

$$K_w = [H^+] \times [OH^-] = (10^{-7}) \times (10^{-7}) = 10^{-14}$$

As a result, the pH and the pOH of pure water both equal 7. Acidic solutions have pH values less than 7 and pOH values greater than 7. Basic (alkaline) solutions have pH values greater than 7 and pOH values less than 7.

When you add an acid or base to an aqueous solution, the concentrations of hydrogen and hydroxide ions shift in proportion so that the following is always true:

pH + pOH = 14

K_a and K_b

As we explain in Chapter 16, the K_a and K_b are equilibrium constants that measure the tendency of weak acids and bases, respectively, to undergo ionization reactions in water.

For the dissociation reaction of a weak acid, HA

HA $\leftrightarrow$ H$^+$ + A$^-$ (also written: HA + H$_2$O $\leftrightarrow$ A$^-$ + H$_3$O$^+$)

the acid dissociation constant, K_a, is

$$K_a = \frac{[H^+][A^-]}{[HA]}$$

Note that H$_3$O$^+$ may be used in place of H$^+$.

For the dissociation reaction of a weak base, A$^-$

A$^-$ + H$_2$O $\leftrightarrow$ HA + OH$^-$

the base dissociation constant, K_b, is

$$K_b = \frac{[HA][OH^-]}{[A^-]}$$

Note that B$^+$ and BOH (base molecule, without or with OH$^-$) may be used in place of HA and A$^-$ (acid molecule, with or without H$^+$).

Moreover, because $K_w = [H^+] \times [OH^-]$, it follows that

$$K_b = \frac{K_w}{K_a}$$

The stronger the acid or base, the larger the value of K_a or K_b, respectively.

It's often useful to consider pK_a or pK_b, the negative log of K_a or K_b:

$pK_a = -\log K_a$

$pK_b = -\log K_b$

The pK_a or pK_b is equivalent to the pH at which half the acid or base has undergone the dissociation reaction. This equivalence is reflected in the Henderson-Hasselbach equation, which can be used to relate pH to the relative concentrations of a weak acid (HA) and its conjugate base (A^-):

$$pH = pK_a + \log\left(\frac{[A^-]}{[HA]}\right)$$

Flip to Chapter 17 for more information about the Henderson-Hasselbach equation.

Chapter 25

Ten Annoying Exceptions to Chemistry Rules

*E*xceptions seem like nature's way of hedging its bets. As such, they are annoying — why can't nature just go ahead and *commit?* But nature knows nothing of our rules, so it certainly isn't going out of its way to annoy you. Nor is it going out of its way to make things easier for you. Either way, seeing as you have to deal with exceptions, we thought we'd corral many of them into this chapter so you can confront them more conveniently.

Hydrogen Isn't an Alkali Metal

In the field of psychology, Maslow's hierarchy of needs declares that people need a sense of belonging. It's a good thing hydrogen isn't a person because it belongs nowhere on the periodic table (which we describe in Chapter 4). Although hydrogen is usually listed atop Group lA along with the alkali metals, it doesn't really fit. Sure, hydrogen can lose an electron to form a +1 cation, just like the alkali metals, but hydrogen can also gain an electron to form hydride, H^-, especially when bonding to metals. Furthermore, hydrogen doesn't have metallic properties, but typically exists as the diatomic gas, H_2.

These differences arise largely from the fact that hydrogen has only a single $1s$ orbital and lacks other, more interior orbitals that could shield the valence electrons from the positive charge of the nucleus. (See Chapter 4 for an introduction to s and other types of orbitals.)

The Octet Rule Isn't Always an Option

REMEMBER

An *octet,* as we explain in Chapter 4, is a full shell of eight valence electrons. The octet rule states that atoms bond with one another to acquire completely filled valence shells that contain eight electrons. It's a pretty good rule. Like most pretty good rules, it has exceptions.

 ✔ Atoms containing only $1s$ electrons simply don't have eight slots to fill. So, hydrogen and helium obey the "duet" rule.

✔ Certain molecules contain an odd number of valence electrons. In these cases, full octets aren't an option. Like people born with an odd number of toes, these molecules may not be entirely happy with the situation, but they deal with it.

✔ Atoms often attempt to fill their valence shells by covalent bonding (see Chapter 5). Each covalent bond adds a shared electron to the shell. But covalent bonding usually requires an atom to donate an electron of its own for sharing within the bond. Some atoms run out of electrons to donate, and therefore can't engage in enough covalent bonds to fill their shell octets. Boron trifluoride, BF_3, is a typical example. The central boron atom of this molecule can only engage in three B–F bonds, and ends up with only six valence electrons. This boron is said to be electron deficient. You might speculate that the fluorine atoms could pitch in a bit and donate some more electrons to boron. But fluorine is highly electronegative and greedily holds fast to its own octets. C'est la vie.

✔ Some atoms take on more than a full octet's worth of electrons. These atoms are said to be *hypervalent* or *hypercoordinated*. The phosphorus of phosphorus pentachloride, PCl_5, is an example. These kinds of situations require an atom from period (row) 3 or higher within the periodic table. The exact reasons for this restriction are still debated. Certainly, the larger atomic size of period 3 and higher atoms allows more room to accommodate the bulk of all the binding partners that distribute around the central atom's valence shell.

Some Electron Configurations Ignore the Orbital Rules

Electrons fill orbitals from lowest energy to highest energy. This is true.

The progression of orbitals from lowest to highest energy is predicted by an Aufbau diagram. This isn't always true. Some atoms possess electron configurations that deviate from the standard rules for filling orbitals from the ground up. For Aufbau's sake, why?

Two conditions typically lead to exceptional electron configurations.

✔ First, successive orbital energies must lie close together, as is the case with $3d$ and $4s$ orbitals, for example.

✔ Second, shifting electrons between these energetically similar orbitals must result in a half-filled or fully filled set of identical orbitals, an energetically happy state of affairs.

Want a couple of examples? Strictly by the rules, chromium should have the following electron configuration:

$$[Ar]3d^4 4s^2$$

Because shifting a single electron from $4s$ to the energetically similar $3d$ level half-fills the $3d$ set, the actual configuration of chromium is

$$[Ar]3d^5 4s^1$$

For similar reasons, the configuration of copper is not the expected $[Ar]3d^9 4s^2$, but instead is $[Ar]3d^{10}4s^1$. Shifting a single electron from $4s$ to $3d$ fills the $3d$ set of orbitals.

Flip to Chapter 4 for full details on electron configurations and Aufbau diagrams.

One Partner in Coordinate Covalent Bonds Giveth Electrons; the Other Taketh

To form a covalent bond (as we explain in Chapter 5), each bonding partner contributes one electron to a two-electron bond, right? Not always. *Coordinate covalent bonds* are particularly common between transition metals (which are mostly listed as group B elements on the periodic table) and partners that possess lone pairs of electrons.

Here's the basic idea: Transition metals have empty valence orbitals. *Lone pairs* are pairs of nonbonding electrons within a single orbital. So, transition metals and lone-pair bearing molecules can engage in Lewis acid-base interactions (see Chapter 16). The lone-pair containing molecule acts as an electron donor (a Lewis base), giving both electrons to a bond with the metal, which acts as an electron acceptor (a Lewis acid). When this occurs, the resulting molecule is called a *coordination complex*.

The partners that bind to the metal are called *ligands*. Coordination complexes are often intensely colored and can have properties that are quite different than those of the free metal.

All Hybridized Orbitals Are Created Equal

Different orbital types have grossly different shapes. Spherical s orbitals look nothing like lobed p orbitals, for example. So, if the valence shell of an atom contains both s- and p-orbital electrons, you might expect those electrons to behave differently when it comes to things like bonding, right? Wrong. If you attempt to assume such a thing, valence bond theory politely taps you on the shoulder to remind you that valence shell electrons occupy hybridized orbitals. These hybridized orbitals (as in sp^3, sp^2, and sp orbitals) reflect a mixture of the properties of the orbitals that make them up, and each of the orbitals is equivalent to the others in the valence shell.

Although this phenomenon represents an exception to the rules, it's somewhat less annoying than other exceptions because hybridization allows for the nicely symmetrical orbital geometries of actual atoms within actual molecules. VSEPR theory presently clears its throat to point out that the negative charge of the electrons within the hybridized orbitals causes those equivalent orbitals to spread as far apart as possible from one another. As a result, the geometry of sp^3-hybridized methane (CH_4), for example, is beautifully tetrahedral.

Check out Chapter 5 for the details on VSEPR theory and hybridization.

Use Caution When Naming Compounds with Transition Metals

The thing about transition metals is that the same transition metal can form cations with different charges. Differently charged metal cations need different names, so chemists don't get any more confused than they already are. These days, you indicate these differences by using Roman numerals within parentheses to denote the positive charge of the metal ion. An older method adds the suffixes *–ous* or *–ic* to indicate the cation with the smaller or larger charge, respectively. For example:

Cu^+ = copper (I) ion or cuprous ion

Cu^{2+} = copper (II) ion or cupric ion

Metal cations team up with nonmetal anions to form ionic compounds. What's more, the ratio of cations to anions within each formula unit depends on the charge assumed by the fickle transition metal. The formula unit as a whole must be electrically neutral. The rules you follow to name an ionic compound must accommodate the whims of transition metals. The system of Roman numerals or suffixes applies in such situations:

$CuCl$ = copper (I) chloride or cuprous chloride

$CuCl_2$ = copper (II) chloride or cupric chloride

Chapter 6 has the full scoop on naming ionic and other types of compounds.

You Must Memorize Polyatomic Ions

Sorry, it's true. Not only are polyatomic ions annoying because you have to memorize them, but they pop up everywhere. If you don't memorize the polyatomic ions, you'll waste time trying to figure out weird (and incorrect) covalent bonding arrangements, when what you're really dealing with is a straightforward ionic compound. Here they are in Table 25-1 (see Chapter 6 for more information):

Table 25-1	Common Polyatomic Ions
–1 Charge	*–2 Charge*
Dihydrogen phosphate ($H_2PO_4^-$)	Hydrogen Phosphate (HPO_4^{2-})
Acetate ($C_2H_3O_2^-$)	Oxalate ($C_2O_4^{2-}$)
Hydrogen Sulfite (HSO_3^-)	Sulfite (SO_3^{2-})
Hydrogen Sulfate (HSO_4^-)	Sulfate (SO_4^{2-})
Hydrogen Carbonate (HCO_3^-)	Carbonate (CO_3^{2-})
Nitrite (NO_2^-)	Chromate (CrO_4^{2-})
Nitrate (NO_3^-)	Dichromate ($Cr_2O_7^{2-}$)

–1 Charge	–2 Charge
Cyanide (CN^-)	Silicate (SiO_3^{2-})
Hydroxide (OH^-)	**–3 Charge**
Permanganate (MnO_4^-)	Phosphite (PO_3^{3-})
Hypochlorite (ClO^-)	Phosphate (PO_4^{3-})
Chlorite (ClO_2^-)	**+1 Charge**
Chlorate (ClO_3^-)	Ammonium (NH_4^+)
Perchlorate (ClO_4^-)	

Liquid Water Is Denser than Ice

Kinetic molecular theory, which we cover in Chapter 10, predicts that adding heat to a collection of particles increases the volume occupied by those particles. Heat-induced changes in volume are particularly evident at phase changes, so liquids tend to be less dense than their solid counterparts. Weird water throws a wet monkey wrench into the works. Because of H_2O's ideal hydrogen-bonding geometry, the lattice geometry of solid water (ice) is very "open" with large empty spaces at the center of a hexagonal ring of water molecules. These empty spaces lead to a lower density of solid water relative to liquid water. So, ice floats in water. Although annoying, this watery exception is quite important for biology.

No Gas Is Truly Ideal

No matter what your misty-eyed grandparents tell you, there were never halcyon Days of Old when all the gases were ideal. To be perfectly frank, not a single gas is really, truly ideal. Some gases just approach the ideal more closely than others. At very high pressures, even gases that normally behave close to the ideal cease to follow the Ideal Gas Laws that we talk about in Chapter 11.

When gases deviate from the ideal, we call them _real gases_. Real gases have properties that are significantly shaped by the volumes of the gas particles and/or by interparticle forces. To account for these nonideal factors, chemists use the van der Waals equation. Compared to the Ideal Gas equation, $PV = nRT$, the following van der Waals equation includes two extra variables, a and b. The variable a corrects for effects due to particle volume. The variable b corrects for interparticle forces. The van der Waals equation is appropriate for gases at very high pressure, low temperature conditions, and when gas particles have strong mutual attraction or repulsion.

$$\left(P + \frac{n^2 a}{V^2}\right)(V - nb) = nRT$$

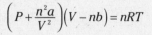

Common Names for Organic Compounds Hearken Back to the Old Days

Serious study of chemistry predates modern systematic methods for naming compounds. As a result, chemists persistently address a large number of common compounds, especially organic compounds, by older, "trivial" names. This practice won't change anytime soon. A cynical take on the situation is to observe that progress occurs one funeral at a time. A less cynical approach involves serenely accepting that which you cannot change, and getting familiar with these old-fashioned names. Table 25-2 lists some important ones; head to Chapter 20 for details on organic compounds.

Table 25-2	Common Names for Organic Compounds	
Formula	*Systematic Name*	*Common Name*
CH_3CO_2H	Ethanoic acid	Acetic acid
CH_3COCH_3	Propanone	Acetone
C_2H_2	Ethyne	Acetylene
$CHCl_3$	Trichloromethane	Chloroform
C_2H_4	Ethene	Ethylene
H_2CO	Methanal	Formaldehyde
CH_2O_2	Methanoic acid	Formic acid
C_3H_8O	Propan-2-ol	Isopropanol

Index

• *H* •

• *I* •